Farida Allal

Thermodynamics of solutions and mixtures

Farida Allal

Thermodynamics of solutions and mixtures

Phase diagrams

ScienciaScripts

Foreword

The thermodynamics of solutions and mixtures studies the energy properties of multiconstituent systems. It is the science that governs energy exchanges between systems undergoing a physical or chemical transformation and the external environment. It finds its applications in the design and modelling of chemical processes, requiring precise knowledge of the thermodynamic properties of pure bodies and mixtures.

Thermodynamics of solutions and mixtures is one of the fundamental disciplines of solution chemistry that deals with acid-base reactions, redox reactions, solubility and complexation equilibria.

This book is about understanding phase diagrams of binary systems, it has three chapters.

The first chapter recalls some basic definitions and relations, such as composition scales and their conversion formulas: molar fraction, molarity, mass fraction. Fugacity, chemical potential and activity are also presented in this chapter.

In chapter II, we present the binary liquid-vapor and liquid-solid phase diagrams.

Chapter III contains the excess quantities and thermodynamic models of activity coefficients.

This book ends with the solutions to the exercises mentioned in the previous chapters.

Chapter I: Fundamentals of Solutions and Mixtures

I.1 Solutions and mixtures

A solution can be liquid or solid. A distinction is made between the solvent: the majority compound and the solute: the minority compound. The solvent is generally a pure compound. It can also be formed by a mixture of determined composition.

In the solutions, we see two scales of composition:

Molar fraction xs of the solvent

Molality $_{mi}$, molarity $_{ci}$ or molar fraction xi for solutes.

A mixture can be a liquid, solid or gas. Non-ionic solutions are almost always treated as mixtures.

The thermodynamic treatment of mixtures is simpler. The scale of composition is the mole fraction. The reference state for the chemical potential of all constituents is the pure body.

I.2 Thermodynamic treatment of mixtures

I.2.1 Partial molar quantities

Let us consider an extensive state function of a thermoelastic system (reversible exchange of elastic energy with the external medium). The quantity Z can be the volume V, the internal energy U, the enthalpy H, the entropy S. The extensive state functions also depend on the composition. The most used variables in chemistry are temperature and pressure. The compositional scale is the number of moles of each constituent nj, hence :

$$\mathbf{Z = Z(T, P, n_i, \dots \dots \dots \dots \dots n_j \dots \dots \dots.)}$$

The partial molar quantity Zj relative to the component j and associated with the extensive property Z is the partial derivative of Z with respect to the number of moles nj, at constant temperature and pressure :

$$z_j = \left(\frac{\partial Z}{\partial n_j}\right)_{T,P,n_i \neq j}$$

At constant temperature and pressure, the extensive quantities obey the following relation:

$$Z_{T,P}(\lambda n_1 \ldots\ldots\ldots\ldots\ldots\ldots . \lambda n_j \ldots\ldots\ldots\ldots\ldots\ldots .) = \lambda Z_{T,P}(n_1, \ldots\ldots\ldots\ldots\ldots . n_j)$$

Example

Consider a mixture containing n1 and n2 moles of components 1 and 2. When the amount of material is doubled, the volume is doubled.

$$V_{T,P}(2n_1, 2n_2) = 2V_{T,P}(n_1, n_2)$$

From the mathematical point of view, the extensive quantities are homogeneous functions of order 1 with respect to the quantity of matter. The application of Euler's identity which governs these functions gives :

$$Z_{T,P}[n_1, \ldots\ldots\ldots\ldots n_j] = n_1 \left(\frac{\partial Z}{\partial n_1}\right)_{T,P,n_{i\neq j}} + \cdots \ldots\ldots\ldots n_j \left(\frac{\partial Z}{\partial n_j}\right)_{T,P,n_{i\neq j}}$$

Or again:

$$Z_{T,P} = \sum_j n_j Z_j$$

At constant temperature and pressure, an extensive state quantity is the sum of the partial molar quantities weighted by the quantity of matter. The preceding relation calls for two remarks:

Partial molar quantities are different from the molar properties of pure substances. Thus, the volume of a mixture of different species is not the same as the sum of the volumes of the pure substances it contains:

$$V_{T,P} = \sum_j n_j v_j \neq \sum_j n_j v_j^*$$

The molar quantities of pure bodies depend only on temperature and pressure, but the partial molar quantities depend on the composition.

$$Z_j^* = Z_j^*(T, P)$$

$$Z_j = Z_j(T, P, n_1 \ldots \ldots n_j \ldots \ldots)$$

I.2.2. Properties of partial molar quantities

Here are some examples for a system evolving at constant x.

Enthalpy $\qquad H = P + P \qquad \leftrightarrow \qquad h_j = u_j + Pv_j$

Free energy $\qquad A = U - TS \qquad \leftrightarrow \quad a_j = u_j - Ts_j$

Free Enthalpy $\quad G = H - TS \qquad \leftrightarrow \quad G_j = h_j - Ts_j$

The differential of the free enthalpy is given by :

$$dG = Vdp - SdT \leftrightarrow dg_j = v_j dP - s_j dT$$

Hence : $\left(\dfrac{\partial g_j}{\partial P}\right)_{T,n_j} = v_j \qquad\qquad \left(\dfrac{\partial g_j}{\partial T}\right)_{P,n_j} = -s_j$

Gibbs-Helmholtz relationship : $\left[\dfrac{\partial}{\partial T}\left(\dfrac{G}{T}\right)\right]_P = -\dfrac{H}{T^2}$ and $\left[\dfrac{\partial}{\partial T}\left(\dfrac{g_j}{T}\right)\right]_{P,n_j} = -\dfrac{h_j}{T^2}$

Maxwell's relationship : $\left(\dfrac{\partial S}{\partial P}\right)_T = -\left(\dfrac{\partial V}{\partial T}\right)_P$ and $\left(\dfrac{\partial s_j}{\partial P}\right)_{T,n_j} = -\left(\dfrac{\partial v_j}{\partial T}\right)_{P,n_j}$

I.2.3 Chemical potential

The chemical potential is the quantity on which the thermodynamic description of multicomponent systems is based.

Consider a closed thermoelastic system at temperature T and pressure P, exchanging only mechanical energy and heat with the external environment.

$$dU = -PdV + TdS + \sum_j \left(\frac{\partial U}{\partial n_j}\right)_{S,V,n_{i \neq j}} dn_j$$

At constant volume and entropy, the change in internal energy is the sum of the contributions of each component, i.e. :

$$d_{S,V}U = \sum_j \left(\frac{\partial U}{\partial n_j}\right)_{S,V,n_{i \neq j}} dn_j$$

The proportionality factor is μj :

$$\mu_j = \left(\frac{\partial U}{\partial n_j}\right)_{S,V,n_{i \neq j}}$$

The chemical potential is an intensive quantity expressed in J/mol, it depends on temperature, pressure and composition.

Free enthalpy plays a major role in the study of chemical equilibria. Its differential is :

$$dG = d(U + PV - TS)$$

$$= -PdV + TdS + SdT - TdS - SdT + PdV + VdP + \sum_j \mu_j dn_j$$

$$dG = VdP - SdT + \sum_j \mu_j dn_j$$

The expression of the differential of the free enthalpy as a function of the variables T, P and the composition x is also expressed by :

$$dG = \left(\frac{\partial G}{\partial T}\right)_{P,n_j} dT + \left(\frac{\partial G}{\partial P}\right)_{T,n_j} dP + \left(\frac{\partial G}{\partial n_j}\right)_{T,P,n_{i \neq j}} dn_j$$

The term-by-term composition of the two forms of the dG differential shows the identity of the chemical potential with the partial molar free enthalpy gj.

$$\mu_j = g_j = \left(\frac{\partial G}{\partial n_j}\right)_{T,P,n_{i\neq j}}$$

The application of Euler's identity to the free enthalpy gives :

$$G_{T,P} = \sum_j n_j g_j$$

The free enthalpy of a mixture at constant temperature and pressure is the quantity-weighted sum of the chemical potential of each component.

As with the free enthalpy, the chemical potential appears in the differentials of the other two fundamental energy functions, namely, the enthalpy H and the Helmholtz energy A.

$$dH = d(U + PV) = TdS + VdP + \sum_j h_j\, dn_j$$

$$dA = d(U - TS) = -SdT - PdV + \sum_j a_j\, dn_j$$

The chemical potential is identified with the following four derivatives:

$$g_j = \left(\frac{\partial U}{\partial n_j}\right)_{S,V,n_{i\neq j}} = \left(\frac{\partial G}{\partial n_j}\right)_{T,P,n_{i\neq j}} = \left(\frac{\partial H}{\partial n_j}\right)_{S,P,n_{i\neq j}} = \left(\frac{\partial A}{\partial n_j}\right)_{T,V,n_{i\neq j}}$$

I.2.4. Multiphase equilibria

Let be a closed multiphase system with no chemical reaction. At thermodynamic equilibrium, each constituent j is present in the various phases α, β, γ, ... ϕ. There may be several solid and liquid phases but there is only one gas phase. The system evolves at constant temperature and pressure. The free enthalpy is minimal at equilibrium, i.e. :

$$dG_{T,P} = \sum_j g_j^\alpha dn_j^\alpha + \sum_j g_j^\beta dn_j^\beta + \sum_j g_j^\gamma dn_j^\gamma + \cdots \ldots \ldots \sum_j g_j^\varphi dn_j^\varphi = 0$$

We now consider the evolution of the system at constant T and V, the free energy is minimal at equilibrium and is given by the expression :

$$dA_{T,V} = \sum_j g_j^\alpha dn_j^\alpha + \sum_j g_j^\beta dn_j^\beta + \sum_j g_j^\gamma dn_j^\gamma + \cdots \ldots \ldots \sum_j g_j^\varphi dn_j^\varphi = 0$$

In conclusion, in a closed multiphase system, the chemical potential of each component is the same in each phase.

I.2.5. Gibbs-Duhem relationship

For a closed system evolving at constant temperature and pressure, the expression of an extensive quantity is such that :

$$Z_{T,P} = \sum_j n_j\, z_j$$

Hence the derivative dZ is :

$$+\sum_j dn_j\, z_j dZ_{T,P} = \sum_j n_j\, dz_j$$

This differential is also by definition equal to :

$$\sum_j dn_j\, z_j dZ_{T,P} =$$

These two expressions are compatible only if :

$$=0\left[\sum_j n_j\, dz_j\right]_{T,P}$$

This is the Gibbs-Duhem relation. It is useful in the case of binary mixtures.

I.3 Thermodynamic treatment of solutions

I.3.1 Apparent molar size

In a solution, the contribution of the solvent to a state quantity corresponds by convention to its properties in the pure state. The specific contributions of the solutes take into account all the mixing quantities. This information leads to the following expression for an extensive state quantity:

$$Z = n_s z_s^* + \sum_i n_i \varphi_i$$

The molar quantities φ_i associated with the solutes are the apparent molar quantities, these quantities are very sensitive to any variation in composition. Their evolution allows a precise determination of the partial molar quantities.

I.3.2 Partial molar size

For a binary solution containing only one solute :

$$Z = n_s z_s^* + n_i \varphi_i$$

The derivation of Z with respect to the number of moles of solute gives the expression of the partial molar size of the solute:

$$Z_i = \left(\frac{\partial Z}{\partial n_i}\right)_{T,P,n_s} = \varphi_i + n_i \left(\frac{\partial \varphi_i}{\partial n_i}\right)_{T,P,n_s} = \varphi_i + \left(\frac{\partial \varphi_i}{\partial \ln n_i}\right)_{T,P,n_s}$$

I.3.3. Infinite dilution

The infinite dilution state of a solution, noted by ∞, is obtained by extrapolation of its properties, when the molar fraction of the solvent tends towards 1.

$$Z_i^\infty = \lim_{Z_s \to 1} Z_i$$

I.4. Composition scales in solutions

- **Molarity**: is the amount of solute per unit mass of solvent.

$$m_i(mol/Kg) = \frac{n_i}{n_s M_s}$$

- **Molar volume concentration**: is the amount of solute per unit volume of solution.

$$C_i (\text{mol/m}^3) = \frac{n_i}{V}$$

The molar concentration by volume and the molarity ci (mol/L) are related by the relation :

$$C_i = c_i * 1000$$

- **The mole fractions of** the solute and the solvent are dimensionless quantities, they are expressed by :

$$x_i = \frac{n_i}{n_s + \sum_i n_i} \quad \text{and} \quad x_i = \frac{n_s}{n_s + \sum_i n_i}$$

- **The mass fraction wi** is the quotient of the mass of solute by the mass of solution. It is directly related to the mass percentage by :

$$W_i\% = 100 w_i$$

I.5. Fugacity

The fugacity is a positive quantity of the same dimension as the pressure, defined in the same way in each state of matter, gaseous, liquid or solid. It depends on temperature and pressure.

The fugacity of a perfect gas is the same as the pressure. The fugacity of a liquid at moderate pressure can often be equated with the saturation vapour pressure at the same temperature.

In a gas mixture, Pj is the partial pressure of component j. In an ideal gas mixture, the fugacity is identified with the partial pressure :

$$f_j^{gp} = P_j = y_j P$$

Where P is the total pressure and yj is the mole fraction of component j in the gas phase.

In a liquid or solid mixture, Pj represents the vapour pressure of component j in the gas phase in equilibrium with the condensed phase.

I.5.1. Fugacity of a gas

The isothermal differential of the chemical potential of a pure perfect gas at temperature T and pressure P is proportional to the logarithmic differential of the pressure :

$$dg_T^{gp} = V^{gp}dP = \frac{RT}{P}dP = RTdLnP$$

By analogy with the perfect gas, the isothermal differential of the chemical potential of a real gas is proportional to the logarithmic differential of its fugacity fg :

$$dg_T^{g} = V^{g}dP = \frac{RT}{Pf}df^{g} = RTdLnf^{g}$$

When the pressure tends to 0, real gases tend to the perfect state. The fugacity of a perfect gas is therefore identical to the pressure:

$$f^{gP} = P$$

I.5.2. Fugacity of a liquid

The fugacity of a liquid of molar volume v is defined like that of a gas from the isothermal differential of the chemical potential :

$$dg_T = VdP = RTdLnf$$

I.5.3 Influence of temperature and pressure on the fugacity of a liquid

a) Influence of temperature

The integration of the previous equation between a pressure low enough for the substance to be considered as a perfect vapour and the effective fugacity leads to the relation :

$$g = g^{gp} + RTLn\frac{f}{P}$$

$$\frac{g - g^{gp}}{T} = Rln\frac{f}{P}$$

The derivation with respect to T of the two members of this last equation leads to the Gibbs-Helmhotz relation:

$$\left[\frac{\partial}{\partial T}\left(\frac{g}{T}\right)_P\right] = -\frac{h}{T^2}$$

b) *Influence of pressure*

The influence of the pressure on the fugacity is given by the expression :

$$\left(\frac{\partial lnf}{\partial P}\right)_T = \frac{v}{RT}$$

Unlike gases, pressure has a relatively small effect on the condensed phases.

Hence $\quad \left(\frac{\partial lnf}{\partial T}\right)_P = -\frac{h - h^{gp}}{RT^2} = \frac{L_v}{RT^2}$

Where Lv is the latent heat of vaporization of a liquid or sublimation of a solid.

I.5.4. *Fugacity in a mixture*

The fugacity of a component in a mixture of any physical state is defined from the isothermal variation of the chemical potential:

$$d_T g_j = v_j dP = RTlnf_j$$

With the limiting condition:

$$\lim_{P \to 0}\left(\phi_i = \frac{f_i}{P} = 1\right)$$

Where Φi is the fugacity coefficient of constituent j.

The integration of the previous equation gives the general expression for the chemical potential of a component in a mixture, i.e. :

$$g_j = g_j^{ref} + RTln\left(\frac{f_j}{f_j^{ref}}\right)$$

and :

$$f_j = f_j^{ref}exp\left(\frac{g_j - g_j^{ref}}{RT}\right)$$

Fugacity is an intrinsic property of a component, directly related to its chemical potential. It depends on the temperature, pressure and composition of the liquid and vapour.

I.5.5. Fugacity and multiphase equilibrium

The equality of chemical potentials

$$g_j^{\alpha} = g_j^{\beta} = \ldots\ldots\ldots g_j^{\varphi}$$

implies that of fugacities:

$$f_j^{\alpha} = f_j^{\beta} = \ldots\ldots\ldots f_j^{\varphi}$$

I.5.6. Calculation of the fugacity of a pure fluid

Let be a pure liquid or gaseous fluid. By subtracting the logarithmic derivatives of the pressure, we obtain the logarithmic differential of the fugacity coefficient ϕi, that is :

$$dln\left(\frac{f}{P}\right) = \left(\frac{v}{RT} - \frac{1}{P}\right)dP$$

I.5.7.1. Fugacity of a gas

We place ourselves at a sufficiently low pressure so that the fluid is in the state of condensed vapor. The integration of the previous equation between P=0 at an effective pressure, leads to the expressions of the fugacity coefficients and the fugacity of the vapor:

$$\varphi^{\text{gaz}} = \exp \int_0^P \left(\frac{v^{\text{gaz}}}{RT} - \frac{1}{P} \right) dP$$

$$f^{\text{gaz}} = P \exp \int_0^P \left(\frac{v^{\text{gaz}}}{RT} - \frac{1}{P} \right) dP$$

I.5.7.2. Fugacity of a liquid

The fugacity in the condensable vapor state is given by the above equation. However, when liquid-vapor equilibrium is established, the fugacities of the liquid and vapor being equal, the fugacity of the boiling liquid is therefore :

$$f^* = P^* \exp \int_0^{P^*} \left(\frac{v^{\text{gaz}}}{RT} - \frac{1}{P} \right) dP$$

Where f* and P* represent, respectively, the fugacity of the boiling liquid and its saturation vapour pressure.

If the vapor is perfect, the fugacity of the boiling liquid is equal to its saturation vapor pressure.

Application exercises

<u>**Exercise 1**</u>

1- Recall the expression for the chemical potential of a component in an ideal liquid mixture.

2- A liquid mixture, assumed to be ideal, of 900 g of water and 80 g of methanol at a temperature of 20 °C is considered. The saturation vapour pressures of water and methanol are 18 and 94 mmHg respectively at 20 °C. Determine:

a- The total pressure of the gas mixture.

b- The mole fraction of the constituents in the gas phase. Conclude.

Exercise 2

The figure below shows the variations in partial pressures P1 and P2 and total pressure P of a liquid mixture of tin tetrachloride SnCl4 (noted A1) and carbon tetrachloride CCl4 (noted A2), as a function of the composition of carbon tetrachloride, x(CCl4), in the liquid at 20°C.

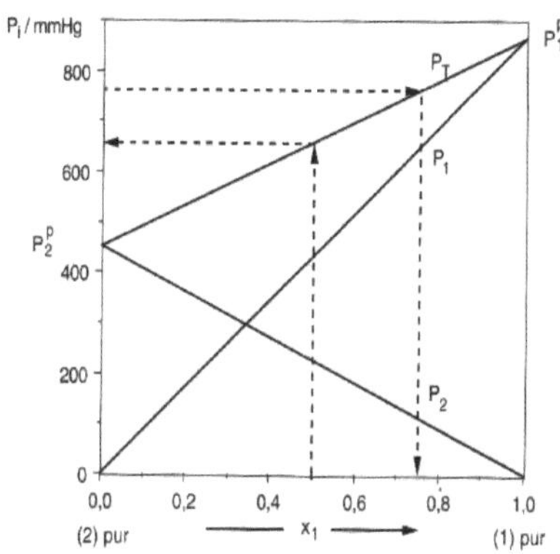

1. Let P_1^* and P_2^* be the saturation vapour pressures of SnCl4 and CCl4 respectively at temperature T. What relationships exist between P1, P2, P_1^* and P_2^* ?

2. Assuming that the gas phase in equilibrium with the liquid mixture can

be considered as perfect and that the partial molar volumes of the pure liquids are negligible compared to those of the gases. v_i^* of the pure liquids are negligible compared to those of the gases, show that the mixture considered is an ideal mixture.

Exercise 3

1. We consider a real mixture with positive deviation.

   a. Draw the diagram showing the partial pressures in the steam

as a function of the mole fraction x2.

   b. For a constituent, with reference to the pure constituent, show that :

$$\gamma_i = \frac{P_i}{P_i^* x_i}$$

   c. For a constituent, referring to the infinitely diluted state of the solution,

show that :

$$\gamma_{i,x} = \frac{P_i}{k_i x_i}$$

Exercise 4

Ethyl bromide $C_2H_4Br_2$ (1) and 1,2-dibromopropane (2) form a series of ideal solutions in the entire compositional range.

At 85°C, the vapour pressures of the two pure liquids are 173 and 127 mmHg respectively.

   a) Plot on the same graph the variations of P1, P_2 and P_T, respectively,

partial pressures of (1), (2) and total pressure, as a function of x1, composition of (1) in the liquid phase at 85°C.

   b) If 10 g of (1) are dissolved in 80 g of (2), calculate the total pressure above the solution at 85°C. It is assumed that the composition of the liquid phase is not changed by the formation of the vapour phase, i.e. the volume of the vapour phase is very small. C=12; H=1; Br=80.

   c) Calculate the composition of the vapour phase in equilibrium with the previous solution.

We will take y1, the composition of (1) in gas phase.

Exercise 5

The mixture of dichlorobenzene (1) and bromobenzene (2) behaves ideally throughout the compositional range. The vapour pressures of the pure components at 137 °C are 863 and 453 mmHg respectively.

1. Express P1, $_{P2}$ and $_{PT}$, respectively partial pressures of (1), (2) and total pressure as a function of x1, molar fraction of (1) in the liquid solution at 137°C.
2. Plot P1, $_{P2}$ and $_{PT \, on}$ the same graph as a function of x1.
3. Determine:
   a. The composition of the liquid mixture which has a normal boiling point of 137°C (boiling under $_{PT=1atm}$).
   b. The composition of the steam in equilibrium with this liquid mixture at its boiling point).
   c. The composition of the vapour and the total pressure above a solution at 137 °C containing an equal number of moles of chlorobenzene and bromobenzene in liquid phase.

Exercise 6

Consider the solutions water (1) + 2-methyl-9-propanol (2) at 30°C. We give :

x2	1.000	0.960	0.925	0.865	0.802	0.700
P2/mmHg	56.9	54.8	53.0	50.1	48.1	46.75
x2	0.656	0.382	0.274	0.202	0.0508	0.018
P2/ mmHg	43.4	36.6	35.0	33.4	28.4	15.2

1. Draw the curve $_{P2=f}$(x2). Does it deviate positively or negatively from the law of ideal solutions?
2. Calculate the activity of (2) taking the pure component (2) at 30 °C as the reference state.
3. Calculate the activity of (2) taking as a reference state the pure component (2) at 30 °C in a hypothetical solution which would follow Henry's law.

Chapter II: Phase diagrams

II.1 Definition

Equilibria between phases are expressed by the equality of the chemical potentials of the different phases in equilibrium, moreover, the chemical potential of a constituent is the same in all phases in equilibrium, i.e. :

$$\mu_i^\alpha = \mu_i^\beta = \,\ldots\ldots\ldots\,\mu_i^\varphi$$

II.2 Phases and constituents

A constituent can be :

- Any atom in the periodic table ;
- A stable molecule such as oxides, carbides and silicates.

A phase can be liquid, solid (crystal or amorphous) or gas.

Solid phases are listed by α, β, γ..... or by the A_xB_y composition.

II.3 Phase rule or variance

The number of intensive variables that must be fixed to determine the state of a chemical system is called the variance of a system.

For a multi-component, multi-phase system, consisting of Nc constituents and Nph in presence, the degree of freedom or variance V is equal to :

$$V = 2 + N_{Ph}(N_c - 1) - N_c\left(N_{ph} - 1\right)$$

$$V = 2 + N_c - N_{ph} = 2 + C - \phi$$

__Examples:__

- ✓ The pressure is fixed, the system is formed by two components and a single phase. The variance is: V=1+2-1=2 (the intensive variables to be fixed are the temperature and the composition)
- ✓ The pressure is fixed, the system is formed of two components and two phases. The variance in this case is equal to 1 (the temperature is fixed and the compositions in the two phases are determined).
- ✓ The pressure is fixed, the system is formed of two components and three phases. The variance is zero. It is an invariant.

II.4. Phase diagrams

Phase diagrams describe a system in equilibrium. They make it possible to predict, for a given mixture, the constitution of the phases present, in equilibrium with each other. They are used in metallurgy, ceramics, and more often for polymers.

II.4.1. Construction of a phase diagram

The curves of the phase diagram determine :

- ➤ the limits of domains in which phases can exist;
- ➤ the composition and proportions of these different phases.

II.4.2. Raoult's and Henry's laws

Consider two distinct chemical systems A_1 and A_2, distributed in two phases, one is liquid and the other is gaseous in equilibrium. Two equations translate these equilibria:

$$\mathbf{A_1(liq) = A_1(vap) \quad et \quad A_2(liq) = A_2(vap)}$$

At a given temperature, the partial pressures of the constituents in the gas phase are represented as a function of the composition of the liquid phase by the following curves:

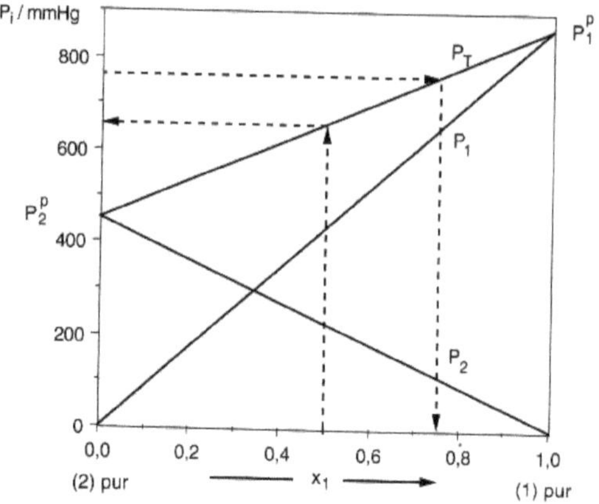

Figure II.1: Total and partial vapour pressures: Case of an ideal solution

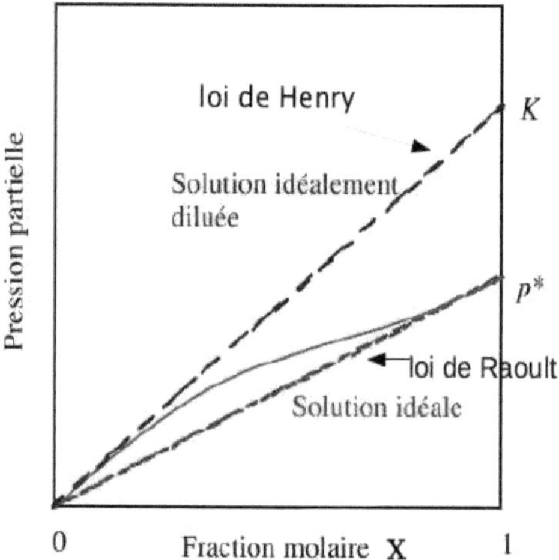

Figure II.2: Total and partial vapour pressures: Case of a real mixture

Case 1: Ideal mixture: in this case, the two curves $P_{i=f}(x_i)$ are straight lines. Such a mixture is ideal. The constituents verifying for any composition, Raoult's law:

$$\mathbf{P_i = x_i P_i^*}$$

Where P_i^* is the saturation vapour pressure of the pure component i.

Such mixtures are obtained with constituents of similar structure.

Case 2: real mixture: in the case of the real mixtures, one observes a strong deviation from the ideality. The curves are not any more straight lines. These deviations are obtained when the mixtures are made up of compounds with different structures. The strong interactions between the molecules are responsible for thermal and volumetric effects at the time of the mixture and of the deviation from the ideality.

Nevertheless, at very low concentrations, the solute obeys Henry's law:

$$\mathbf{P_i = x_i^l k_i}$$

ki is Henry's constant, homogeneous at one pressure.

If $k > P_i^*$ the deviation is positive and the interactions in pure molecules are more important.

If $k < P_i^*$ the deviation is negative and the interactions in the mixture are more important than in the pure molecules.

II.5. Liquid-vapour equilibrium diagrams

II.5.1. Liquid-vapour equilibrium of an ideal binary system. Total miscibility in the liquid state

The study of the liquid-vapour equilibrium of a binary system of two pure bodies A_1 and A_2, assumed to be miscible in all proportions in the liquid state, takes into consideration the following quantities:

x_i^V : molar fraction of Ai in the steam ;

x_i^L : molar fraction of Ai in the liquid ;

P_i : partial pressure of Ai in the steam ;

$P_{si}(T)$ saturation vapour pressure of Ai at temperature ;

μ_i^V chemical potential of Ai in the steam ;

μ_i^L chemical potential of Ai in the liquid.

The system is divariant (v=2), there must exist for a given temperature, a relation between the pressure and the composition of the two constituents in the liquid on the one hand, and in the vapour on the other hand.

Ideal solution: Raoult's law

A liquid solution is said to be ideal if the chemical potential of each component is of the form :

$$\mu_i^L = \mu_i^{0L} + RT \ln x_i^L$$

The expression for the chemical potential in the vapor phase is :

$$\mu_i^V = \mu_i^0(T) + RT \ln P_i$$

We distinguish two types of diagrams: isothermal and isobaric.

II.5.1.1. Isothermal diagram

The diagram is obtained by representing PT, P_1^*, P_2^* as a function of x_i^L or x_i^V we obtain two curves framing a domain :

> ➢ A boiling curve, which gives $P_{total} = f(x_i^L)$. This curve is a straight line which

is obtained from Raoult's law:

$$P_T = P_1^*(x_1^L) + P_2^*\left(1 - (x_1^L)\right)$$

Or :

$$P_T = P_2^* + x_1^L(P_1^* - P_2^*)$$

> A dew curve in the form of a hyperbola, giving $P_{total} = f(x_i^V)$ and

deduced from Dalton's law:

$$x_1^V = \frac{P_1^*}{P_1^* - P_2^*} * \frac{P_T - P_2^*}{P_T}$$

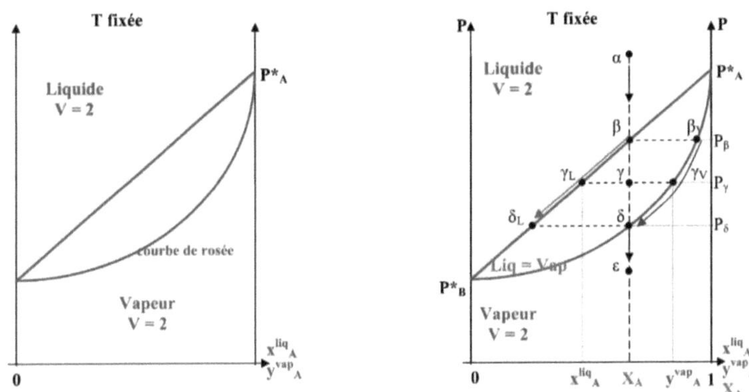

II.5.1.2 Physical interpretation of the boiling and dew point curve.

Reading the diagram

Let's study the isothermal compression of a gas mixture A-B.

The representative point of the system is ε.

- ✓ From ε to δ: the representative point of the system moves on the vertical of abscissa XA. The system remains single-phase and the composition of the system does not change.
- ✓ In δ: the [1st] drop of liquid appears, its mole fraction in A is given by the point δ. The mole fraction in A of the vapor is : $x_A^V = y_A^V = X_A$.
- ✓ From δ to β: compression is continued, the system is two-phase. During this stage, the amount of liquid increases at the expense of the amount of gas. The liquid and the vapour become enriched in A, which is the most volatile

compound, and therefore the one that will be liquefied with the greatest difficulty.

✓ From β to α: the system is single-phase liquid, its composition no longer evolves, the representative point moving on the vertical of abscissa XA. The composition is the same as that of the initial vapor.

II.5.1.3. Isobaric equilibrium diagram

As the temperature changes, the spindle formed by the dew and boiling curves creates a tube-like surface. The boiling curve is below the dew curve.

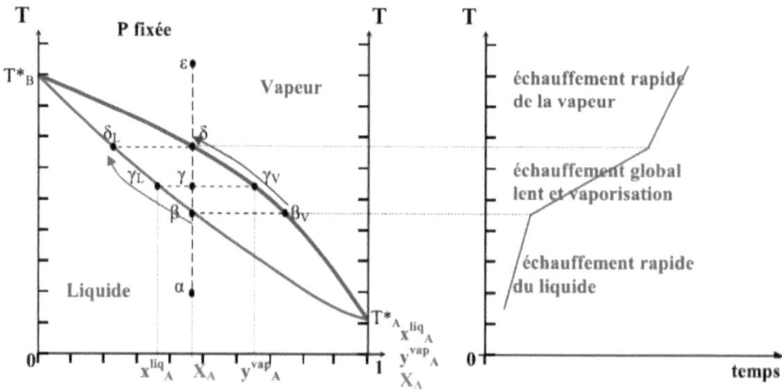

Figure II.3: Construction of a phase diagram. a) Binary phase equilibrium diagram with total miscibility in the solid state. b) Thermal analysis curve.

II.5.1.4. Reading the diagram

✓ Starting from the point α in the domain of the liquid (low temperature) initially assumed to be poor in A. By increasing the temperature, the liquid starts boiling at the point β (appearance of the [1st] bubble of the vapor) belonging to the boiling curve. This first bubble has the composition $(X_\beta)_v$, therefore richer in the most volatile constituent (A).

✓ If you continue heating, the figure point will be in the two-phase zone (L+V).

✓ The vaporization of the liquid will be completed when the composition of the vapor phase

will be the same as the starting liquid (δ point).

II.5.1.5. Chemical moment theorem

The practical purpose of the chemical moment theorem is the knowledge of the ratio of the numbers of moles of liquid and vapor in equilibrium for a certain temperature.

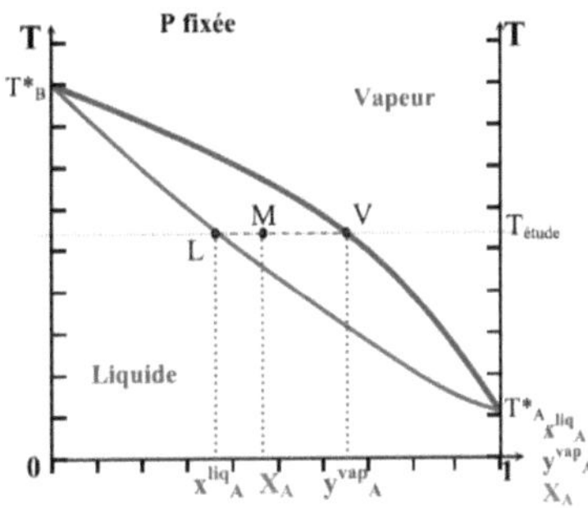

Let us consider the point M belonging to the spindle delimited by the two dew and boiling curves. It is therefore in the two-phase system. This mixture was obtained by heating the initial liquid mixture, of composition XA.

Initially :$(n_A)_0 = (x^L_A)_0 (n_L)_0$ and $(n_B)_0 = (x^L_B)_0 (n_L)_0$

Where $(n_L)_0$ total number of moles of liquid (1+2)

A TM: the two-phase system comprises: nL moles of liquid and nV moles of vapor, knowing that :

$$n_L + n_V = (n_L)_0$$

In the liquid and vapor phase, the mole numbers of component A are given by :

$$n^L_A = x^L_A n_L \text{ and } n^V_A = x^V_A n_V$$

By conservation of matter for A, we get :

$$n_A^L + n_A^V = (n_A)_0 = (x_A^L)_0 (n_L)_0 = (x_A^L)_0 (n_L + n_V)$$

$$x_A^L n_L + x_A^V n_L = (x_A^L)_0 (n_L + n_V)$$

$$[x_A^V - (x_A^L)_0] n_V = [(x_A^L)_0 - x_A^V] n_L$$

Hence:

$$\frac{n_V}{n_L} = \frac{[(x_A^L)_0 - x_A^V]}{[x_A^V - (x_A^L)_0]} = \frac{ML}{VM}$$

II.5.1.6. Chemical potential of component Ai for an ideal binary mixture.

Let the following physico-chemical equilibrium apply:

Solution liquide binaire idéale : $(A_1(L) + A_2(L)$ $\Longleftrightarrow$ Vapeur saturante (mélange de deux gaz parfaits)

This equilibrium results in the equality of the chemical potentials in the two phases, i.e. :

$$\mu_i^L = \mu_i^V$$

$$\mu_i^V = \mu_i^{0V}(T, P) + RT \ln \frac{P_i}{P^0} = \mu_i^0(T, P) + RT \ln \frac{x_i^L P_i^*}{P^0}$$

Where μ_i^{0V} is the reference chemical potential

For an ideal liquid solution, the chemical potential is written :

$$\mu_i^L = \mu_i^{0L}(T) + RT \ln x_i^L$$

μ_i^{0L} is the reference chemical potential of the pure component Ai at the reference temperature and pressure. However, the pressure has little influence on the chemical potential of the condensed phase, since : $\left(\frac{\partial \mu_i}{\partial P}\right)_T = \bar{V}_{M_i}$ (low partial molar volume) and we can take :

$$\mu_i^0(T, P) = \mu_i^0(T)$$

Raoult's law is therefore of the form :

$$\left(\mu_i^0\right)_L + RT\ln x_i^L = (\mu_i^0)_V + RT\ln\frac{P_i}{P^0}$$

II.5.2 Liquid-vapour equilibrium of a real binary system with total miscibility in liquid phase

We consider binary liquid solutions whose properties deviate from ideal solutions.

II.5.2.1. Deviations from ideality: isothermal diagram P(x) at temperature T

For many liquid mixtures, Raoult's laws are no longer followed over a large range of mole fractions. The chemical potentials will therefore admit more complicated expressions.

For diluted solutions, simplified expressions can be found, from which Raoult's law can be verified in this area:$x_i^L \to 0$.

We necessarily introduce the notion of activity ai in place of composition xi.

For $(Ai)_L$: $\mu_i^L = \mu_i^{*L} + RT\ln a_i$ with $a_i = \gamma_i x_i$

At the physico-chemical equilibrium of Ai : $\mu_i^L = \mu_i^V$

$$\mu_i^{*L} + RT\ln a_i = \mu_i^{*V} + RT\ln\frac{P_i}{P^0}$$

Where μ_i^{*L} and μ_i^{*V} are the reference states of the pure component Ai.

II.5.2.2. Activities of the constituents Ai for the real solutions

1. In the vicinity of xi=1, the component Ai is considered to be the solvent, which

which allows to express the equality of the chemical potentials in the two phases as follows:

$$\mu_i^{0L} + RT\ln a_i = \mu_i^{0V} + RT\ln \frac{P_i}{P^0}$$

With : $P_i = x_i^V P_T$ (Dalton relationship)

If component Ai is considered to be alone in the liquid phase : $a_i=1$ and $P_i=P_i^*$

$$\mu_i^{0V} + RT\ln \frac{P_i^*}{P^0} = \mu_i^{0L} + RT\ln 1$$

Hence: $a_i = \dfrac{P_i}{P_i^*}$

This expression allows the determination of the activity of the component Ai by measuring the partial pressures P_i and the saturation vapour pressure P_i^*.

❖ When $x_i^L \to 1$ Raoult's law holds, then $\gamma_i \to 1$ and $a_i = x_i^L$

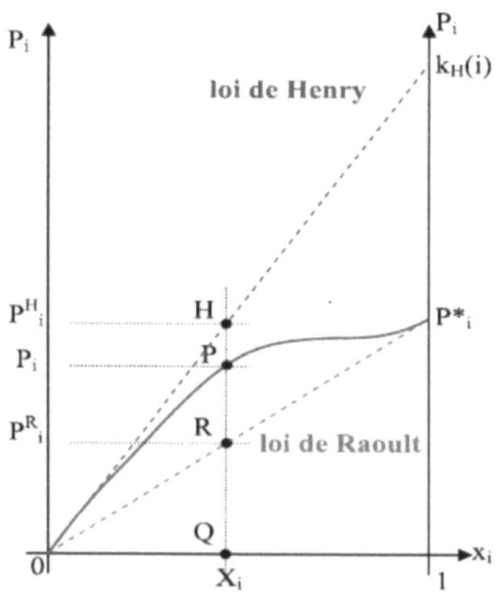

Figure II.4: Activities of real solutions

In conclusion:

$$\mu_i^L = \mu_i^{0L} + RT\ln a_i = \mu_i^{0L} + RT\ln \gamma_i x_i$$

2. In the vicinity of xi→0, the component Ai is considered to be the solute, the solution is therefore real and diluted.

We can observe on the diagram of figure II.4, that the real partial pressure P_i of Ai in the solute state deviates strongly from the pressure expected by Raoult.

For xi<<1, we accept the linear law : $(\mathbf{P_i})_{\text{réelle}} = \mathbf{k_{H,i} x_i^L}$

Where $k_{H,i}$ is Henry's constant, it is homogeneous to a pressure.

Henry's law can be applied in the composition domain: $0 < x_i^L < 0,15$

> When $k_{H,i} > P_i^*$: The deviation from Raoult's law is positive.

> When $k_{H,i} < P_i^*$: The deviation from Raoult's law is negative.

The chemical potential for solute Ai when $x_i^l \to 0$ is written :

$$\mu_i^L = \mu_i^{0L} + RT\ln a_i = \mu_i^{0V} + RT\ln \frac{P_i}{P^0}$$

And

$$\mu_i^V = \mu_i^{0V} + RT\ln \frac{k_{H,i} x_i^L}{P^0}$$

Hence:

$$\mu_i^L = \mu_i^{0V} + RT\ln \frac{k_{H,i}}{P^0} + RT\ln x_i^L = \mu_i^{*L} + RT\ln x_i^L$$

With μ_i^{*L} reference chemical potential in the standard state when $x_i^L \to 0$.

- **<u>General form of the chemical potential in a real solution</u>**

Let: a_i: the activity of Ai in the solution, the reference being the pure body;

a_i^{∞} the activity of Ai in the solution, the reference being the infinitely diluted solution.

For the activity :

$$a_i \rightarrow P_i = a_i P_i^* = \gamma_i x_i P_i^*$$

$$a_i^\infty \rightarrow P_i = a_{i,\infty}' P_i^* = \gamma_i^\infty x_i k_{H,i}$$

For the chemical potential in the vapor phase :

$$\mu_i^V = \mu_i^{0V} + RT\ln\frac{P_i^*}{P^0} + RT\ln a_i$$

$$\mu_i^V = \mu_i^{0V} + RT\ln\frac{k_{H,i}}{P^0} + RT\ln a_{i,\infty}'$$

In liquid phase :

$$\mu_i^L = \mu_i^{0L} + RT\ln\ln a_i$$

$$\mu_i^V = \mu_i^{0V} + RT\ln a_{i,\infty}'$$

II.5.2.3. Isobaric or non-ideal isothermal diagram

As a first approximation, we can assimilate the vapour phase to a perfect gas, but the solution is not ideal. The partial pressures in the vapour phase are not proportional to the molar fractions in the liquid phase.

The liquid-vapour equilibrium is considered as a static equilibrium, obtained when the number of molecules of each body which pass per unit of time, an element of the separation surface from the liquid to the vapour is equal to the number of these molecules, which pass from the vapour to the liquid.

Two morphologies are possible:

- ✓ The diagram deforms slightly (the spindle widens).
- ✓ An azeotropic point Z appears.

Example 1: Quasi-ideal mixture water (1) + methanol (2)

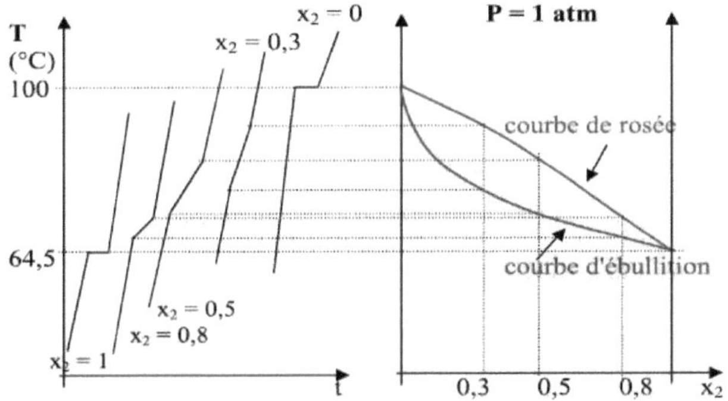

Figure II.5: Construction of a water-methanol isobaric diagram

Example 2: Water (1) + ammonia (2)

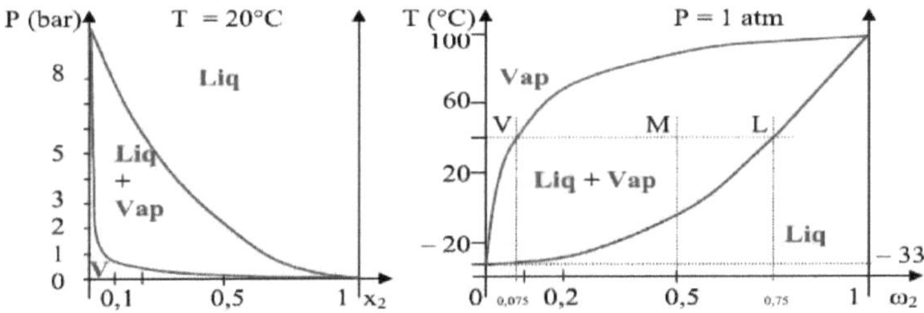

Figure II.6: Isobaric and isothermal water-ammonia binary diagrams

b) Real mixtures with azeotrope formation

The phenomenon of azeotropy appears when the temperature versus composition curve (at constant pressure) of the system reaches an extremum.

In the case of non-ideal solutions, the boiling and dew curves are modified. At constant T and for strong positive deviations from ideality, the total vapor pressure curve passes through a maximum. The temperature varies in the opposite direction to the vapour pressure: a maximum vapour pressure is associated with a minimum boiling temperature.

For negative deviations from ideality, an azeotrope occurs at minimum vapour pressure and maximum boiling temperature. The azeotropic composition is the same for the liquid and for the vapour: xi=yi.

As $x_i = y_i$ we have $\gamma_i(az) = P_i/P_i^* \, x_i = P_i/P_i^* \, y_i$ and as $y_i = P_i/P_T$ hence :

$$\gamma_i(az) = P_T/P_i^*$$

An azeotrope is a homogeneous liquid mixture of two components that boils at a constant temperature; the vapour and the liquid then have the same composition.

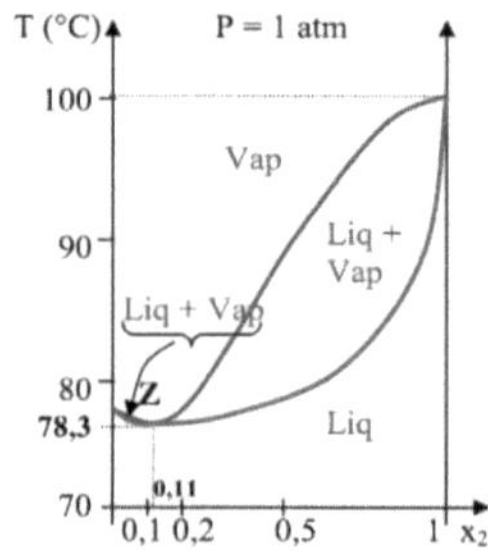

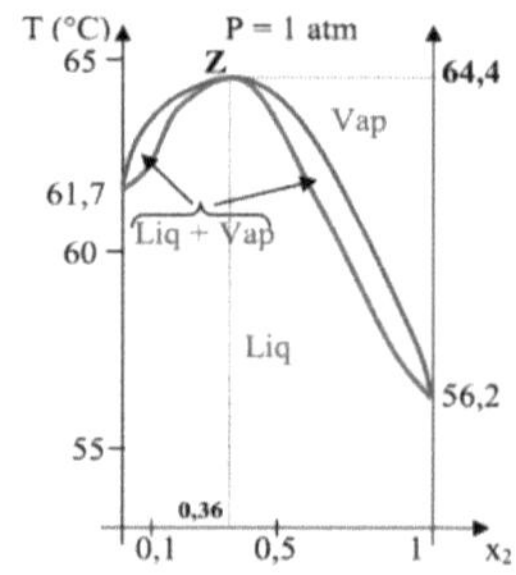

Figure II.7: Isobaric binary diagrams with presence of a minimum azeotrope: ethanol-water mixture and a positive azeotrope: chloroform-acetone mixture

➤ Case of positive deviation (positive homoazeotropy) :

P$_{total}$ (real non-ideal mixture>Pideal predicted by Raoult

➤ Case of negative deviation (negative homoazeotropy):

P$_{total}$ (real non-ideal mixture < Raoult's predicted ideal

c) GIBBS-KONOVALOV theorem. Azeotropic point:

Azeotropic mixture.

➤ An extremum of total pressure P$_T$ at constant temperature leads to the equality of the compositions of the liquid and vapor phases:

$$x_i^L = x_i^V$$

> If the isothermal diagram shows a maximum azeotrope, the isobaric diagram shows a minimum and vice versa.

> An azeotropic mixture behaves like a pure body.

> All mixtures such that $xi \neq xZ$ are zeotropic.

>

II.5.3. Non-ideal liquid-vapour equilibrium diagram with partial miscibility in liquid phase.

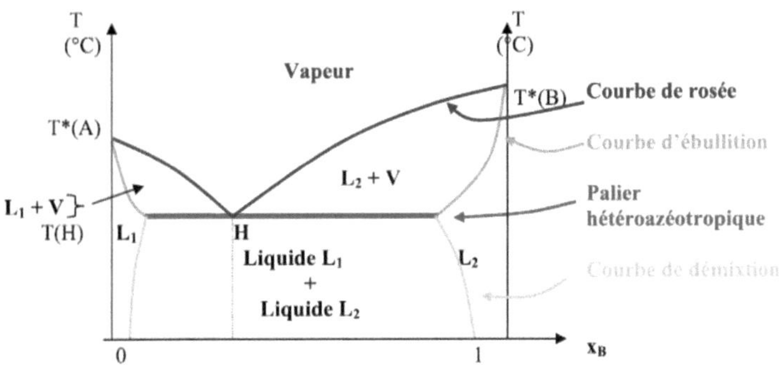

Figure II.8: Non-ideal liquid-vapour equilibrium diagram with partial miscibility in liquid phase

II.5.4. Non-ideal liquid-vapour equilibrium diagram with immiscibility in the liquid phase.

When the two liquids are immiscible, each liquid is pure in its phase. Consequently, and at the change of state, if we start from a mixture of two liquids, we will have two liquid phases and a gaseous phase, i.e. a variance v=1. The binary diagram thus constructed, presents a heteroazeotrope as shown in the figure below:

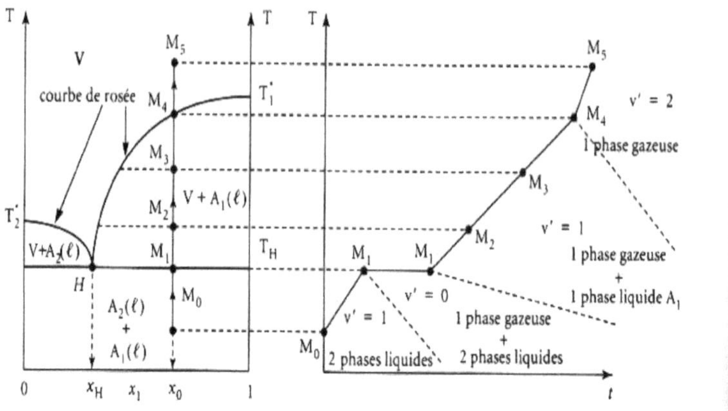

Figure II.9: Binary diagram and thermal analysis of a mixture with zero miscibility in the liquid state

II.6. Liquid-solid equilibrium diagrams of binary mixtures

II.6.1 Introduction

A phase diagram makes it possible to predict, for a given mixture, the constitution of the phases present, in equilibrium, with each other. Two physical factors influence the nature and composition of the phases present:

1. Temperature plays an important role in the modification of the mechanical properties of solids.
2. The pressure that is usually neglected in condensed systems.

Liquid-solid equilibria play a very important role in chemistry, they establish the existence of phases and specify their conditions and fields of existence, they are the basis of many methods of treatment, preparation, separation and purification.

The calculation of phase equilibria of binary systems is based on the concept of chemical potential.

At equilibrium, the chemical potentials of each component are, like temperature and pressure, equal in all phases present.

Solubility depends not only on the activity coefficient of the solute, which is a function of the intermolecular forces between the solute and the solvent, but also on the fugacity of the standard state and the fugacity of the pure solid.

The equilibrium equation is written as :

$$f_2\,[\text{solide pur}] = f_2\,(\text{soluté en solution liquide}) = \gamma_2 x_2 f_2^0$$

Where: x_2 Is the solubility (composition) of the solute in the solvent.

γ_2 is the activity coefficient of the liquid phase.

f_2^0 is the standard state fugacity.

From equation (II.1):

$$x_2 = f_2\,(\text{solide pur})/\gamma_2 f_2^0$$

Thus, solubility depends on the ratio of fugacities as shown in the previous equation.

The choice of the standard state is arbitrary. It is defined as the subcooled liquid at the solution temperature.

II.6.2. Equations of the liquidus and solidus curves

Liquidus equation : $T = f(x_i^L)$

Solidus equation : $T = f(x_i^S)$

In the solid state, A and B are immiscible, hence the chemical potential :

$$\mu_A^S = \mu_B^S = \mu_i^S = \left(\mu_i^{0S}\right)_T$$

The liquid solution is assumed to be perfect, so we have for component i :

$$\mu_i^L = \mu_i^{0L}(T) + RT\ln x_i^L$$

At equilibrium, the chemical potentials in the different phases are equal, thus :

$$\mu_i^S = \mu_i^L = \left(\mu_i^{0S}\right)_T = \mu_i^{0L}(T) + RT\ln x_i^L$$

Now for: Ai(Solid)→Ai(liquid), we have the variation of free enthalpy:

$$\Delta G_{T_{fusion}}^0 = \mu_i^{0L} - \mu_i^{0S} = -RT\ln x_i^L$$

Applying the Gibbs-Helmholtz relation, we get :

$$\frac{\partial}{\partial T}\left(\frac{\partial G_i^0}{T}\right) = -\frac{\Delta H_{fusion(i)}^0}{T^2} = -R\left[\frac{d\ln\left(x_i^L\right)}{dT}\right]$$

Hence:

$$\ln\left(x_i^L\right) = \frac{\left(\Delta H_{fusion}^0\right)_i}{R}\left[\frac{1}{(T_{fusion})_i} - \frac{1}{T}\right]$$

The above equation is applicable for both constituents of the binary mixture, i.e. :

$$\ln(x_1^L) = \frac{\left(\Delta H_{fusion}^0\right)_1}{R}\left[\frac{1}{(T_{fusion})_1} - \frac{1}{T}\right]$$

$$\ln(x_2^L) = \frac{\left(\Delta H_{fusion}^0\right)_2}{R}\left[\frac{1}{(T_{fusion})_2} - \frac{1}{T}\right]$$

The solidus curve corresponds to :

$x_1^S = 1\ et\ x_2^S = 0$ for component A

$x_2^S = 1\ et\ x_1^S = 0$ for component B

For a real liquid solution, the equations become :

$$\ln(\gamma_1 x_1^L) = \frac{\left(\Delta H_{fusion}^0\right)_1}{R}\left[\frac{1}{(T_{fusion})_1} - \frac{1}{T}\right]$$

$$\ln(\gamma_2 x_2^L) = \frac{\left(\Delta H_{fusion}^0\right)_2}{R}\left[\frac{1}{(T_{fusion})_2} - \frac{1}{T}\right]$$

II.6.3. Solid state miscibility rules

For two solids to be fully miscible in the solid state, they must have sufficient analogy :

- Same crystal structure;
- Nearby atomic rays;
- Equal valences;
- Similar electronegativities.

II.6.4. Case of binary mixtures with total miscibility in the solid state

There is total reciprocal solubility, both in the liquid and solid state. In the solid state, a single solid solution is obtained by total miscibility.

The two constituents of the binary mixture must have the same crystallographic type and mesh size. This type of diagram is rarely found in practice.

Example: [Cu, Ni], [Ag, Au], [Au, Pt], [Sn, Bi], [Sb, Bi]

a) Isobaric diagram :

Let's take the case of the Copper-Nickel binary (figure II.9)

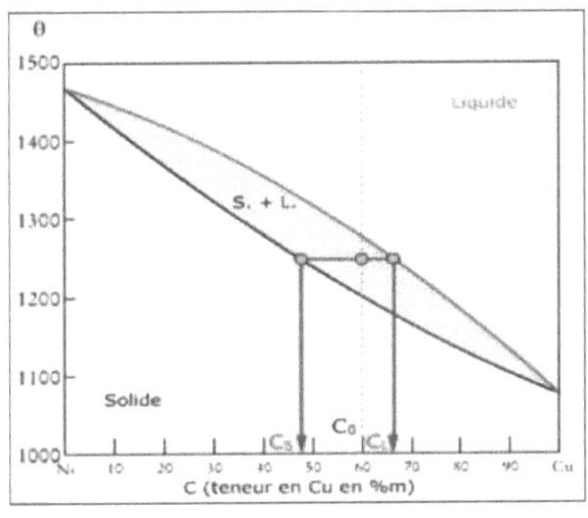

Figure II.10: Ni-Cu diagram with total miscibility in solid phase

The diagram has two curves:

- The liquidus: this is the crystallization curve. It gives the temperature of crystallization starting when the liquid is cooled (appearance of the first crystals of solid solution).
- The solidus: this is the melting curve. It gives the starting melting temperature of a solid solution that is heated (appearance of the first drops of liquid phase.

If one of the rules is not respected, we speak of partial miscibility in the solid state between A and B. Indeed, the addition of atoms from B to A, or vice versa, leads to a distortion of the network of atoms from A and an increase in the internal energy of the system. The laws of thermodynamics then lead the mixture to separate into two phases, one rich in A, the other rich in B, or to form intermediate compounds defined by AxBy.

II.6.5. Partial miscibility diagram in the solid state

In this type of diagram, two kinds of solidus solutions can be observed:

-α: primary solid solution of B in A (rich in A)

-β: primary solid solution of A in B (rich in B).

The two solidification spindles are connected in the central region of the diagram, with the appearance of an invariant equilibrium point between a common liquid phase and two solid phases. Two types of diagrams can be distinguished, depending on the characteristic temperature of the remarkable triple point, in relation to the melting temperatures of the pure constituents.

II.6.5.1. Diagram with formation of a eutectic mixture

Characteristics of the eutectic

- The eutectic is characterized by the reaction : $L \leftrightarrow A_{sol} + B_{sol}$.
- On solidification, a single liquid phase gives rise to two solid phases.
- The composition of the eutectic is fixed.

➢ The temperature of the eutectic is lower than the melting temperature of the two pure bodies.

The following physical equilibria are generated:

$$A_1(liq) \leftrightarrow A_1(sol)$$

$$A_2(liq) \leftrightarrow A_2(sol)$$

With a variance: V=2+1-ϕ. The pressure has little effect on these condensed phases.

If two solids A and B are in equilibrium with the liquid, V=0, we have a triple point, defined at a given temperature and composition: this is the eutectic symbolized by E.

In conclusion:

Eutectic diagrams are characterized by the presence of a demixing zone and two solidification spindles connecting at a eutectic point E (figure II.10).

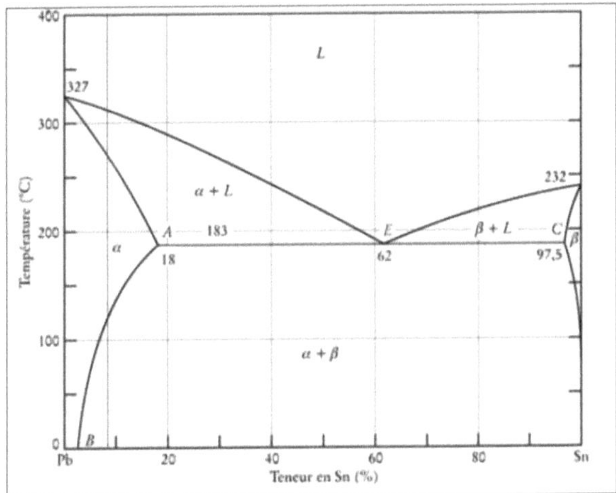

Figure II.11: Liquid-solid equilibrium diagram at partial miscibility: case of Pb-Sn alloy

The curve AB in the figure represents the solubility limit of tin in lead and the curve CD represents the solubility limit of lead in tin. These two curves constitute the solvent lines. The solubility of one element in the other varies with temperature. Thus, the solubility of tin in lead goes from 18% at 183°C to 2% at room temperature.

Thermal diagram of crystallization

From the figure below, we see that these two-phase equilibria form two groups: the first is that of a liquid solution in equilibrium with an A-rich solid solution. It corresponds to the liquidus $\theta_{FA}E$ and the solidus $\theta_{FA}I$. Point I represents the solid solution with 80% of A and 20% of B.

The second is that of liquid solutions in equilibrium with a solid solution rich in B. It corresponds to the liquidus $\theta_{FB}E$ and the solidus $\theta_{FB}J$. The point J represents the solid solution with about 80% of B and 20% of A.

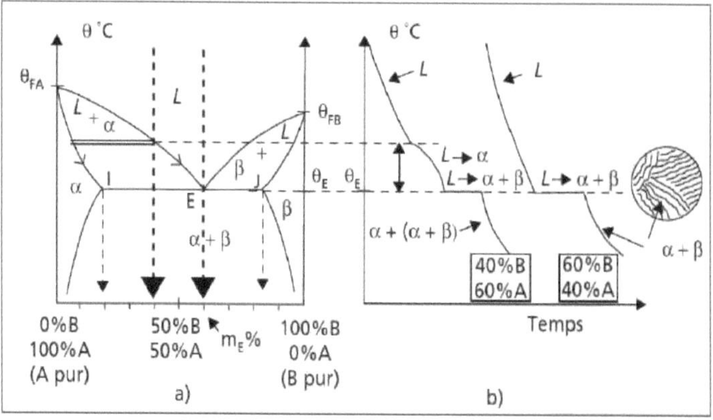

Figure II.12: Construction of a phase equilibrium diagram. a)- Binary phase equilibrium diagram A-B with eutectic reaction. b)- Thermal analysis curves of any mixture and the eutectic mixture

The thermal analysis of the diagram in the figure allows to follow the cooling of the liquid mixture as a function of time. However:

The cooling of a liquid to 40% in B goes through the following steps:

➢ Cooling until the first crystals of pure A appear.

➢ Crystallization at variable temperature of A, while the liquid is enriched in B.

➢ With further cooling, the eutectic temperature T_E is reached, corresponding to the appearance of the second solid phase. The crystals of A and B form simultaneously at a constant temperature T_E.

➢ At T<<TE, the last drop of the liquid crystallizes, giving rise to the two solids A and B, separated into two phases.

II.6.5.2. Diagrams with eutectoid point

The mechanism of the eutectoid transformation is the transformation of a solid phase into two new solid phases according to the equilibrium :

$$\gamma \leftrightarrow \alpha + \beta$$

The iron-carbon phase diagram is an example of this transformation.

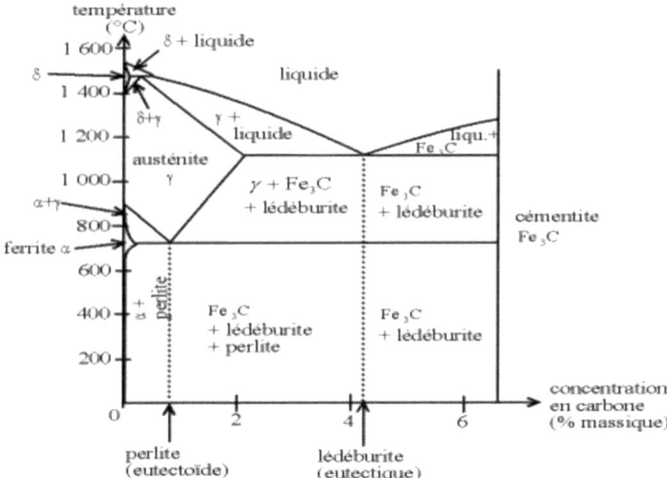

Figure II.13: Liquid-solid equilibrium diagram of the binary mixture Fe-C

II.6.5.3. Diagram with formation of a peritectic mixture

In a peritectic transformation, a liquid phase and a solid phase are transformed into a single solid phase of defined composition.

The peritectic transformation is characterized by:

> balance : $L + \alpha \leftrightarrow \beta$
> During solidification, a new phase is created.
> The composition of the peritectic is fixed.
> The peritectic temperature is between the melting temperatures of the two pure bodies.

The figure shows the liquid-solid equilibrium diagram of the Pb-Pt alloy.

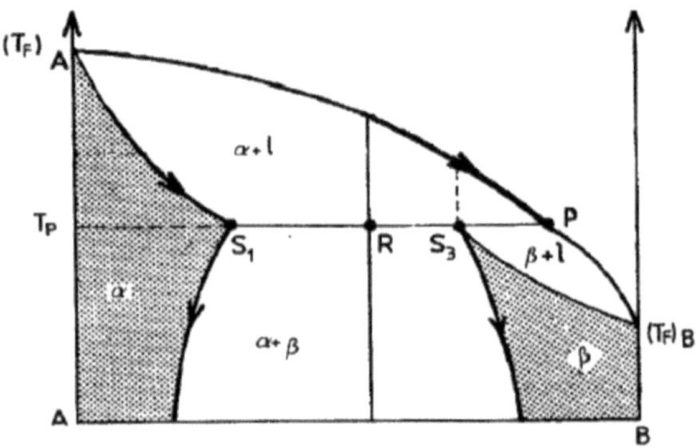

Figure II.14: Equilibrium diagram of Ag-Pt binary phase showing a peritectic reaction

II.6.5.4. Diagrams with peritectoid point

The mechanism of the peritectoid transformation resembles the peritectic transformation, but during this solid transformation, two solid phases are transformed into a new solid phase according to the equilibrium : $\alpha + \beta \leftrightarrow \gamma$

II.6.5.5. Diagrams with defined compounds

It is possible that bodies A and B form a stoichiometric solid of the formula A_mB_n. Two situations are possible:

1. *Defined congruent melting compounds*

In particular proportions, A and B react with each other to form a defined compound A_mB_n. This is shown in the diagram by the formation of a vertical straight line, starting from the composition corresponding to the defined compound.

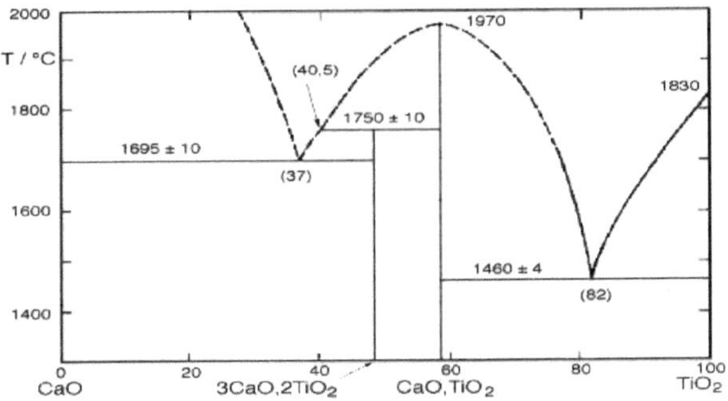

Figure II.15: Phase diagram of the CaO-TiO2 binary system with the presence of defined congruent melting compound (CaO,TiO2)

2. *Non-congruent melting compounds*

The defined compounds formed in specific proportions decompose before reaching their theoretical melting point.

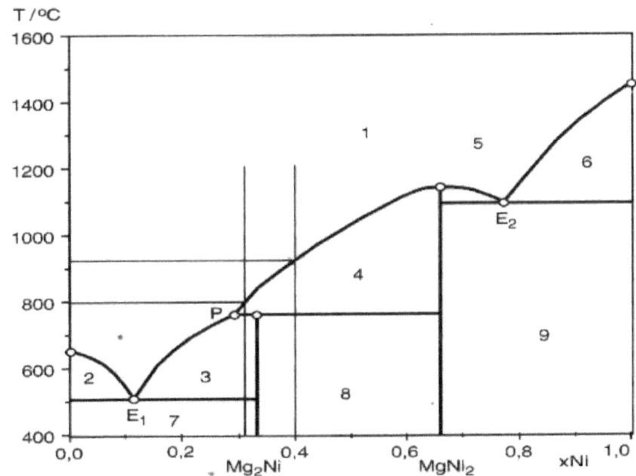

Application exercises

Exercise 1

The partial pressures of the mixtures of water (1) and propanol (2), measured at 25°C, are shown as a function of the mole fraction $x2$ of propanol in the following table:

$x2$	0	0.02	0.05	0.10	0.20	0.40	0.60	0.80	0.90	0.95	1
P2/mmHg	0.0	5.05	10.8	12.7	13.6	14.2	15.5	17.8	19.4	20.8	21.8
P1/mmHg	23.8	23.5	23.2	22.7	21.8	21.7	19.9	13.4	8.1	4.2	0.0

1)- Draw the diagram giving :

The variation of the total pressure P_T as a function of $x2$;

The variations of the vapour pressures $P1$ and $P2 = f(x2)$;

Comment on these diagrams.

2)- a) discuss the shape of the above curves:

in the neighbourhood of $x2=0$; - in the neighbourhood of $x2=1$.

b)- calculate the activity coefficients $\gamma1$ and $\gamma2$, of the constituents of the mixtures for $x2=0.2$, referred to the pure constituents.

(c) calculate the activity coefficients $\gamma_1^\infty, \gamma_2^\infty$ of the constituents for $x2=0.2$, referred to the infinitely diluted constituents.

3)- A mixture corresponding to a molar fraction of propanol of 0.90 is evaporated at 25°C until the composition of the liquid is equal to 0.95. Determine the composition of the vapour during this operation.

Exercise 2

a) The boiling diagram of the system (nitric acid, water) under normal pressure (P=1.013

bar) is given by the following graph, on which the mass fraction w_{HNO3} of the mixture in nitric acid is plotted on the abscissa from left to right.

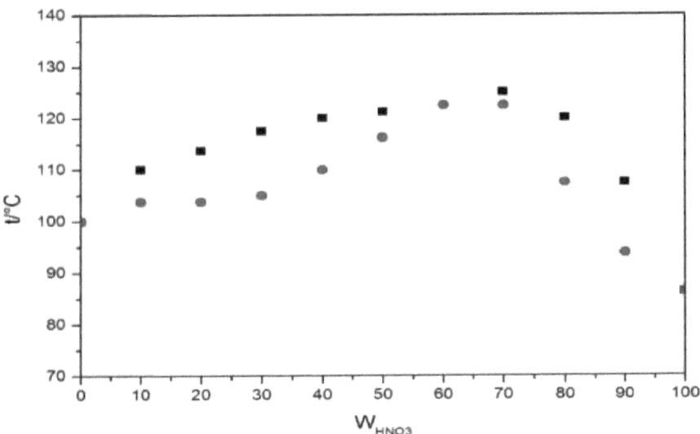

To which phases do the various plane areas in the diagram correspond?

What are the names of the two curves?

What is the name of the liquid mixture whose composition corresponds to the abscissa of the maximum?

What property does such a mixture have?

b) The operations carried out during the rest of the problem will be performed under a pressure of 1.013 bar.

A sample of the mixture obtained by the industrial preparation of nitric acid consists of a total of 4 moles; it contains n1=0.3 moles of nitric acid.

Calculate the total mass fraction w_{HNO3} of nitric acid and show that at temperature $t_I=100°C$ the system is a homogeneous liquid.

Molar masses: Nitric acid: $M1=63$ g/mol; Water: $M2=18$ g/mol

At what temperature t must the sample be brought to the boil?

c) When operating in a closed system, the system is heated up to the temperature t'=110 °C.

When the diagram has a mass fraction on the abscissa, the chemical moment rule applies to the masses of the two phases (whereas if a molar fraction is given on the abscissa, the rule applies to the quantities of matter of the phases).

What are the mass m'_L of the liquid phase and its mass fraction W_{LHNO3} of nitric acid at temperature t'?

Exercise 3

1)- a) Construct, using the following information, the solid-liquid equilibrium diagram of the naphthalene-α-naphthol mixture, made under a pressure of 760 mmHg of mercury.

The table below gives the temperature of the beginning of crystallization of the liquid phase, homogeneous, during cooling, for various mixtures; the compositions are expressed in molar fractions of naphthalene.

Start of crystallization temperature/°C	Composition of the mixture expressed as mole fraction of naphthalene
86	0.25
68	0.50
71	0.75

Melting temperature of pure naphthalene: 80°C

Melting temperature of pure α-naphthol: 96°C

Melting temperature of the eutectic mixture : 61°C

Molar composition of the eutectic mixture: naphthalene 60%, α-naphthol 40%.

Molar mass of naphthalene: 128 g/mol

Molar mass of α-naphthol: 144 g/mol

b) Specify for each area of the diagram, the nature of the phases present.

2)- 50 grams of a liquid mixture, initially taken at 100°C, consisting of 40 g of α-naphthol and 10 g of naphthalene, is slowly cooled.

(a) At what temperature, as read on the diagram, do the first crystals appear? What is their nature?

(b) Give the shape of the cooling curve for this mixture (curve giving the temperature of the mixture as a function of time).

c) Calculate the mass of each phase using the diagram:

When the mixture is cooled to 75°C

when the mixture is cooled to 50 °C.

Exercise 4

In this exercise a part of the Fe-C diagram is presented, which has many applications in the production of cast iron and steel.

Iron has three stable allotropic varieties before melting at atmospheric pressure (Tf=1536°C): T<910°C: Fe α C.C. structure.

910<T<1410°C : Fe γ CFC structure

1410<T<1536°C Fe δ C.C. structure

In the presence of carbon, and depending on its content, iron is likely to form solid solutions.

Consider the Iron-Carbon diagram shown below

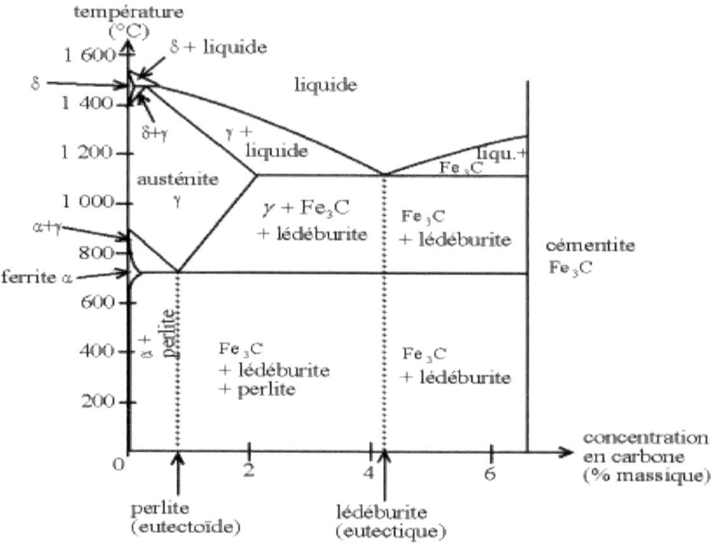

1 - What is this binary diagram, the iron-graphite diagram or the iron-cementite diagram?

2- What does the transformation line at 1495°C represent?

3 - What does the transformation line at 723°C represent?

4 - Indicate the type of alloy containing 0.8% carbon ?

    a) What is the first solid phase that appears at this temperature?

    b) What is the melting point of this alloy?

5- Indicate the nature of the different phases in each area of the diagram.

6- At what temperature is the solubility of carbon in iron maximum in the γ phase?

7- What is the maximum percentage of carbon in α ferrite?

8- A 722-ε°C, what are the proportions of the phases formed by a 1% carbon alloy?

Exercise 5

Consider the binary diagram AgNO3-NaNO3 at constant P.

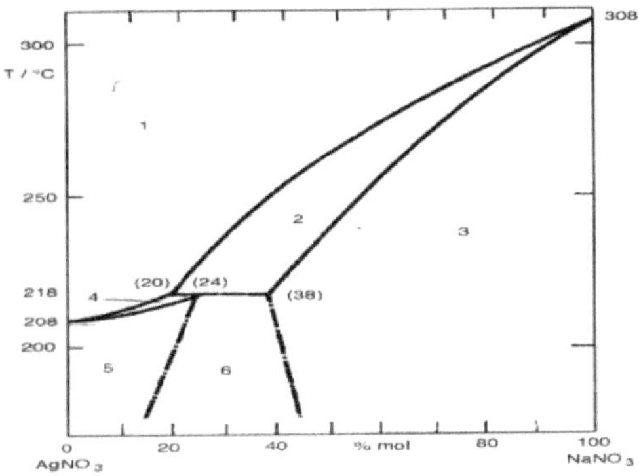

1. Indicate the nature of the phases found in the areas noted 1 to 6.

2. Describe what happens when the binaries are cooled to 22.30 and 60 mol Na NO3 respectively. For each, give the shape of the cooling thermograms and indicate for each portion, the nature of the phases present as well as the evolution of their composition, the reactions if any.

Exercise 6

Consider the CaO-TiO2 binary diagram at constant P.

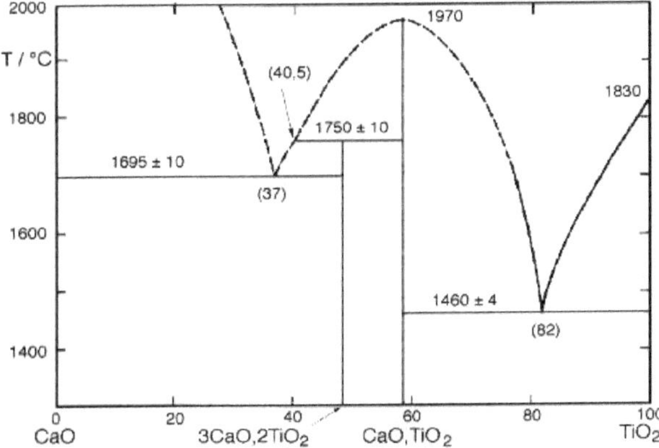

1. Indicate the nature of the phases found in the areas marked 1 to 9.

2. Describe what happens when the binaries are cooled to 45.55 and 70% weight respectively. of TiO2.

3. Give for each the cooling thermograms and indicate for each portion, the phases present as well as the evolution of their composition, the reactions if any.

4. Indicate approximately for the 45% binary the composition of the phases present and their respective quantities at 1800, 1733 and 1600°C.

Exercise 7

Consider the binary A-B at P const. shown below:

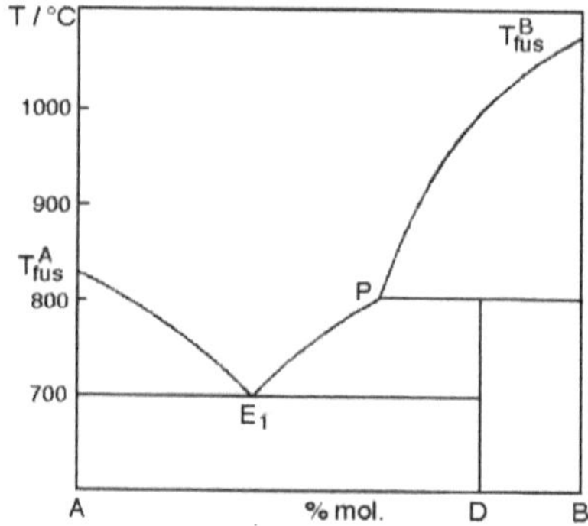

1) Indicate the phases present in the different regions of the diagram.

2) Describe the cooling of a liquid binary mixture with 70% mol of B. Show the thermogram.

3) Give the formula for compound D.

Exercise 8

Below is the lead-bismuth (Pb-Bi) binary equilibrium diagram

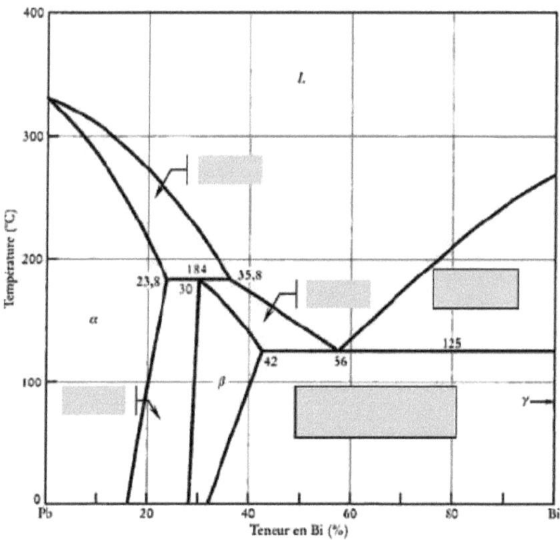

1. On the Pb-Bi diagram complete and calculate the variance for each domain.

2. Give the names and coordinates of the isothermal reactions in the diagram.

3. Determine the chemical composition of the two phases in equilibrium at T=126°C for the 80% Bi alloy.

4. Estimate the mass fractions of the equilibrium phases present at T=126°C, for an alloy of 80% Bi mass.

5. For the 50%-mass Bi alloy, what is the temperature above which the alloy is completely liquid?

6. For the same alloy, at what temperature will it completely solidify?

7. Either the eutectic alloy.

    a. What is its chemical composition?

    b. What phases are present at 95°C?

    c. What is the chemical composition of each of these phases?

    d. Estimate the mass fraction of each of these phases.

8) Check the right answer :

The Pb-Bi binary equilibrium diagram is :

Fully miscible in the liquid state.

A partial miscibility in the solid state.

Not miscible in the solid state.

9) Plot thermal analysis curves for alloys of composition 10%, 20%, 30% and 56% Bi, from 350°C to room temperature.

Exercise 9

We propose to draw the liquid-vapour equilibrium diagram of the H_2O-HNO_3 system, under a pressure of 1 atm. To do this, we have the experimental data summarized in the following table:

Mass fraction of HNO_3	0	0.1	0.15	0.2	0.3	0.4	0.5	0.6	0.66	0.7	0.75	0.8	0.85	0.9	0.95	1
T°C	100	101	102	103	105	109	115	120	123	120	114	105	99	93	88	83
T°C	100	106	109	112	116	118	120	122	123	122	120	117	112	105	96	83
	100								123							83

1. Draw the phase diagram of the binary system H_2O-HNO_3.
2. Indicate for each area the phases present (on the diagram).
3. Indicate the dew point curve on the figure.
4. What is the name of the mixture whose composition corresponds to the maximum temperature?
5. Consider a new mixture M obtained during the preparation of nitric acid. It consists of a total of 100 moles and its molar fraction of HNO_3 is $x_{HNO_3}=0.11$.
   a. Show that the total mass of this mixture M is equal to 2295 g.
   b. Calculate the mass fraction of HNO_3 in this mixture M.
6. At what temperature does this mixture M start to boil?
7. This mixture M is maintained at 109°C.
   a. Give the mass fraction of the phases present.
   b. Calculate the masses of each phase.

Chapter III: Excess quantities and thermodynamic models

III.1 Excess quantities

Excess quantities are of great importance in thermodynamics, both in theory and in practice. They are used in the design of separation processes and are a test for the development of liquid mixture theories.

The excess quantities allow us to predict the nature and intensity of the molecular interactions involved, which result from the combination of the various energy contributions.

The excess quantities are extensive properties, allowing to describe the deviations from the ideality, whose sign and intensity depend on the interactions between the various constituents of the mixture.

A positive deviation is due to the breakup of the associated structures into fragments or molecules.

A negative deviation is due to physical, chemical or geometric interactions.

This allows us to define the excess volume, the excess enthalpy, the excess free enthalpy and the excess entropy:

$$V^E = V - V^{id}$$

$$H^E = H - H^{id}$$

$$G^E = G - G^{id}$$

$$S^E = S - S^{id}$$

There are relationships between these extensive quantities, between the thermodynamic properties from which they are derived, for example:

$$G^E = H^E - TS^E$$

$$G^E = A^E + PV^E$$

The thermodynamic properties correspond to partial molar quantities; in particular, the chemical potential of excess given by the equation:

$$\mu_i^E = \left(\frac{\partial G^E}{\partial n_i}\right)_{T,P,n_j} = \mu_i - \mu_i^{id}$$

Due to the difficulties encountered in experimental measurements of multicomponent systems, semi-empirical and empirical equations are developed to estimate the excess quantities of multicomponent systems.

III.2 Thermodynamic models for estimating activity coefficients

Deviations from ideality are due to short and long range molecular interactions. These deviations are estimated by excess quantities (excess volume, excess enthalpy, excess entropy, excess free enthalpy), summing all contributions.

The activity coefficients of species i in aqueous solutions can only be estimated from models that take into account all interactions, whether physical or chemical.

III.2.1 Physical interaction models

Different physical interaction models have been developed to estimate the enthalpy

and activity coefficients, in non-ideal liquid mixtures, as a function of composition and temperature.

Physical interaction models fall into two categories: semi-predictive and predictive models.

III.2.1.1. Semi-predictive models

III.2.1.1.1. The regular solutions model

a) Van Laar model

Van Laar, proposed a very simple model, he considered a binary mixture consisting of n1 moles of liquid 1 and n2 moles of liquid 2. The mixture is made at constant temperature and pressure. In his model he applied the Van Der Waals equation of state to predict the volumetric properties of pure bodies and mixtures, relying on the following approximations:

- $V^E = 0$
- $S^E = 0$

Thus at constant pressure :

$$g^E = U^E + PV^E - TS^E$$

According to Van Laar's approximations, this results in:

$$g^E = U^E$$

According to the Van Laar model, the free energy is represented by the following expression:

$$g^E = \frac{x_1 x_2 b_1 b_2}{x_1 b_1 + x_2 b_2}\left(\frac{\sqrt{a_1}}{b_1} - \frac{\sqrt{a_2}}{b_2}\right)^2$$

A particular mathematical representation of the previous equation allows us to derive the expressions below:

$$\ln\gamma_1 = \frac{A'}{\left[1 + \frac{A' x_1}{B' x_2}\right]^2}$$

$$\ln\gamma_1 = \frac{B'}{\left[1 + \frac{B' x_2}{A' x_1}\right]^2}$$

$$A' = \frac{b_1}{RT}\left[\frac{\sqrt{a_1}}{b_1} - \frac{\sqrt{a_2}}{b_2}\right]^2$$

$$B' = \frac{b_2}{RT}\left[\frac{\sqrt{a_1}}{b_1} - \frac{\sqrt{a_2}}{b_2}\right]^2$$

Van Laar's equations relate the activity coefficient to the temperature, composition and properties of the pure components (a1, b1) and (a2, b2).

Two important characteristics are deduced from the Van Laar equations:

1. The logarithm of the activity coefficients is inversely proportional to temperature.

2. According to Van Laar's theory, the activity coefficients of each component are greater than 1. This theory always predicts positive deviations from ideality.

b) Scatchard Hildebrand's theory

Scatchard and Hildebrand assumed that Van Laar's theory can be improved, if it will be freed from the approximations of the Van der Waals equation; this can be achieved by defining a parameter C according to :

$$C = \frac{\Delta U_{vap}}{v^L}$$

Where ΔU_{vap} is the vaporization energy. The parameter C is called "cohesive energy density".

In order to complete their theory, Scatchard and Hildebrand assumed that the mixing of the constituents takes place without excess entropy ($S^E = 0$) at constant temperature and pressure ($V^E = 0$), in other words $g^E = U^E$.

Thus, according to this theory, the activity coefficients are defined by :

$$RTln\gamma_1 = v_1 \Phi_2^2 [\delta_1 - \delta_2]^2$$

$$RTln\gamma_2 = v_2 \Phi_1^2 [\delta_1 - \delta_2]^2$$

Where Φi is the volume fraction of component i :

$$\Phi_i = \frac{x_i v_i}{\sum_j x_j v_j}$$

δ is called the solubility parameter and is given by the expression :

$$\delta_i = C_i^{1/2} = \left[\frac{\Delta U_{vap,i}}{v}\right]^{1/2}$$

The equations correspond to a regular solution, this one presents a positive deviation to the ideality.

The solubility parameters δ_1 and δ_2 are temperature dependent, however, their difference is often almost independent of temperature.

The Scatchard Hildebrand theory allows a better representation of the activity coefficients for solutions containing non-polar compounds and for which the deviation from ideality is small.

The regular solution models (Van Laar, Scatchard Hildebrand ...) are not applicable to mixtures containing polar compounds, polymers or hydrogen bonding associations.

III.2.1.1.2. Flory and Huggins theory

According to Flory and Huggins, the deviations from ideality are entropic in origin and correspond to the possibilities of distribution of polymer molecules in a three-dimensional network. A polymer molecule is assimilated to a chain containing a large number of segments of the same dimension, each segment occupying a site in the network.

If we have n1 moles of solvent, n2 moles of polymer and m segments, the total number of sites is $(n1+mn2)_{NA}$

The volume fractions of the solvent and the polymer are given by :

$$\phi_1 = \frac{n_1}{n_1 + mn_2} \quad \text{et} \quad \phi_2 = \frac{mn_2}{n_1 + mn_2}$$

Flory and Huggins showed that, if the amorphous polymer and the solvent are mixed without any thermal effect (athermal behavior), the variation of the free enthalpy and the entropy of mixing are given by the following expression :

$$-\Delta G_{\text{mélange}}/RT = \Delta S_{\text{mélange}}/R = -n_1 \ln \phi_1 - n_2 \ln \phi_2$$

The expression of Flory and Huggins leads to:

$$\frac{S^E}{R} = -x_1 \ln\left[1 - \phi_2\left(1 - \frac{1}{m}\right)\right] - x_2 \ln[m - \phi_2(m - 1)]$$

Therefore, the Flory and Huggins model predicts negative deviations from ideality for an athermal solution:

$$G^E/RT = H^E/RT - S^E/R = 0 - S^E/R < 0$$

$$\ln\gamma_2 = \ln\left[1 - \left(1 - \frac{1}{m}\right)\phi_2\right] + \left(1 - \frac{1}{m}\right)\phi_2$$

The theoretical result of Flory and Huggins can be applied to non-thermal solutions, if and only if a semi-empirical term is involved in the expression of the heat of mixing, which takes into account the energetic interactions between the constituents of the mixture:

$$\frac{\Delta G_{\text{mélange}}}{RT} = n_1\ln\phi_1 + n_2\ln\phi_2 + \chi\phi_1\phi_2(n_1 + mn_2)$$

Where χ is the Flory interaction parameter, it is related to the molecular interaction energies by the inter-exchange parameter w12 :

$$\chi = \frac{Z(2\varepsilon_{21} - \varepsilon_{11} - \varepsilon_{22})}{2kT} = \frac{w_{12}}{kT}$$

The model of Flory and Huggins can be related to the theory of regular solutions by writing :

$$\chi = \frac{v_1}{RT}(\delta_1 - \delta_2)^2$$

Where v_1 is the molar volume of the solvent, δ_1 and δ_2 the solubility parameters of the solvent and the polymer.

III.2.1.1.3. Local composition models

1. Wilson's model

Since its introduction in 1964, the Wilson equation has received much attention, given its ability to accommodate non-ideal but miscible systems.

Wilson's semi-theoretical model is based on the concept of the local volume fraction, which incorporates the effects of the difference in the sizes of the molecules in solution, and on Flory Huggins' theory. It requires only two adjustable parameters per binary system, applies correctly to fully miscible mixtures and to liquid-vapor equilibria.

The expression of the activity coefficient is written as follows:

$$\mathbf{Ln\gamma_i} = -\mathbf{Ln}\left(\sum_{i=1}^{n} x_i \Lambda_{ij}\right) + 1 - \sum_{k=1}^{n} \frac{x_k \Lambda_{ki}}{\sum_{j=1}^{n} x_j \Lambda_{kj}}$$

The binary interaction parameters are defined as follows:

$$\Lambda = \frac{v_j^{0L}}{v_i^{0L}} \exp\left(-\frac{\lambda_{ij} - \lambda_{ii}}{RT}\right)$$

$$\lambda_{ii} = \lambda_{jj} = 1$$

λ_{ij} is the empirically determined interaction energy term. It characterizes the interactions between molecules " i " and " j ".

$$\lambda_{ij} = \lambda_{ji}$$

$$\lambda_{ii} = -\beta(\Delta Hv_i - RT)$$

ΔHv_i being the enthalpy of vaporization of the pure component "i" at a given temperature T and β a proportionality factor.

v_i^{0L} represents the molar volume of the liquid of the pure component i.

Wilson's formulation is written as follows:

$$\Lambda_{ij} = \Lambda_{ij}^0 + \Lambda_{ij}^T T$$

This formulation allows the determination of temperature-dependent interaction parameters.

Λ_{ij}^0, Λ_{ij}^T being the binary interaction coefficients.

Due to its formulation, Wilson's model does not allow to represent binary systems that form a demix.

2. Non-Random Two Liquid (NRTL) model

The NRTL (Non-Random Two Liquid) model, an equation developed by Renon and Prausnitz, is also based on the concept of local composition, but has the advantage over the Wilson model of being able to represent liquid-liquid equilibria.

Indeed, the NRTL model introduces the concept of local composition, based on the hypothesis of the non-random distribution of molecules j around molecule i

. This activity coefficient model can be applied to miscible systems, as well as to partially miscible systems. The calculation of the activity coefficient is given by the equation :

$$\mathbf{Ln\gamma_i} = \frac{\sum_j^n \tau_{ji} G_{ji} x_j}{\sum_k^n G_{ki} x_k} - \sum_j^n \frac{x_j G_{ij}}{\sum_k^n G_{kj} x_k} \left[\frac{\sum_j^n x_k \tau_{kj} G_{kj}}{\sum_k^n G_{kj} x_k} \right]$$

The interaction parameter τ_{ij} of the binary mixture i, j is defined by the following relation:

$$\tau_{ij} = \frac{g_{ij} - g_{jj}}{RT}$$

Where g_{ij} represents a free enthalpy parameter related to the interaction between molecules i and j.

With :

$$\tau_{ii} = \tau_{jj} = 0$$

$$\alpha_{ii} = \alpha_{jj} = 0$$

$$g_{ij} - g_{jj} = C_{ij}^0 + C_{ij}^T (T - 273.15)$$

$$\alpha_{ij} = \alpha_{ij}^0 + \alpha_{ij}^T (T - 273.15)$$

C_{ij}^0, C_{ij}^T, α_{ij}^0 et α_{ij}^T being binary interaction parameters.

The expression for G_{ij} is written as follows:

$$G_{ij} = \exp(-\alpha_{ij}\tau_{ij})$$

With : $G_{ii} = G_{jj} = 1$

Like the Wilson equation, the NRTL model has an important property, only the parameters of binary systems are used to represent multicomponent mixtures.

The NRTL model requires three temperature-dependent parameters per binary system.

3. UNIQUAC (Universal Quasi Chemical) model

The UNIQUAC (Universal Quasi Chemical) model, proposed by Abrams, Prausnitz and Anderson, is based on the concept of local composition. $\mathbf{Ln\gamma_i^{Comb}} = \mathbf{Ln}\frac{\phi_i}{x_i} + \frac{Z}{2}\mathbf{q_i Ln}\frac{\theta_i}{\phi_i} + \mathbf{l_i} - \frac{\phi_i}{x_i}\sum_j^n \mathbf{x_j\, l_j}$ It is based on the concept of local composition and is developed on a theoretical basis of statistical thermodynamics, provided by the Guggenheim quasi chemical network model. The UNIQUAC model requires two adjustable parameters τ_{ij} and τ_{ji} per binary.

The expression of the activity coefficient, shows two terms: A combinatorial term and a residual term.

$$\mathbf{Ln\gamma_i} = \mathbf{Ln\gamma_i^{Comb}} + \mathbf{Ln\gamma_i^{Rés}}$$

$$\mathbf{Ln\gamma_i^{Rés}} = \mathbf{q_i} - \mathbf{q_i Ln}\left[\sum_{j=1}^n \theta_j\,\tau_{ji}\right] - \mathbf{q_i}\sum_{j=1}^n \frac{\theta_j'\tau_{ij}}{\sum_{k=1}^n \theta_k\tau_{kj}}$$

With :

$$\mathbf{l_i} = \frac{Z}{2}(\mathbf{r_i} - \mathbf{q_i}) - (\mathbf{r_i} - \mathbf{1})$$

Z : the coordination number, it is fixed at 10, x_i the molar fraction of component i, ϕ_i: the volume fraction of component i, θ_i: the surface fraction of component i.

The expression for τ_{ij} is written as follows:

$$\tau_{ij} = \exp\left(-\frac{u_{ij}-u_{jj}}{RT}\right)$$

With $\tau_{ii} = \tau_{jj} = 1$

$$\phi_i = \frac{r_i x_i}{\sum_{j=1}^{n} r_j x_j}$$

$$\theta_i = \frac{q_i x_i}{\sum_{j=1}^{n} q_j x_j}$$

r_i and q_i are respectively volume and surface parameters of component i, they are calculated additively by the Bondi method. As for the interaction parameters τ_{ij}, are evaluated by correlations of experimental data.

The wording being:

$$\Delta u_{ij} = \left(u_{ij} - u_{ii}\right)^0 + \left(u_{ij} - u_{ii}\right)^T . T$$

III.2.1.1.4. Predictive models of group contribution

Generally speaking, the activity coefficient is given by a relation of the form :

$$RTLn\gamma_i = RTLn\gamma_i^* + \sum_s v_{si}\left(Ln\Gamma_s - Ln\Gamma_s^{(i)}\right)$$ Where γ_i^* is the combinatorial contribution, Γ_s is the activity coefficient of group s, and v_{si} is the number of groups on molecule i.

Several approaches can be distinguished according to the expression of Γ_s.

Where γ_i^* is the combinatorial contribution, Γ_s is the activity coefficient of group s, and v_{si} is the number of groups on molecule i.

Several approaches can be distinguished according to the expression of Γ_s.

1. Universal Functions Activity Coefficient (UNIFAC) model

The UNIFAC group contribution model is frequently used in the chemical industry for the prediction of liquid phase activity coefficients of multiconstituent systems.

The UNIFAC model is based on the UNIQUAC model and uses the concept of group contributions.

The activity coefficient of component i in a multicomponent mixture, as in the UNIQUAC model, is separated into two terms: a combinatorial term associated with differences in molecular size and shape and a residual term, emanating from interactions between molecular structural groups.

The activity coefficient is given by :

$$\ln\gamma_i = \ln\gamma_i^{Comb} + \ln\gamma_i^{Rés}$$

Several versions have been proposed, depending on the expressions of the activity coefficient of the combinatorial term and that of the residual term

a) **Original UNIFAC**

The combinatorial part of the model is given by the following expression:

$$\mathbf{Ln\gamma_i^{Comb} = Ln\frac{\phi_i}{x_i} + \frac{Z}{2}q_i Ln\frac{\theta_i}{\phi_i} + l_i - \frac{\phi_i}{x_i}\sum_j^n x_j\, l_j}$$

Where ϕ_i , θ_i ri and qi are respectively: the volume fraction, the surface fraction, the relative volume and the relative surface of the component i; they are defined by :

$$\phi_i = x_i r_i / \sum_j x_j r_j$$

$$\theta_i = q_i x_i / \sum_j x_j q_j$$

$$r_i = \sum_k v_{ki} R_k$$

$$q_i = \sum_k v_{ki} Q_k$$

Where k is the total number of groups that make up the molecule, $v_{k,i}$ is the number of

groups of type k in molecule i.

The residual term is expressed by the following equation:

$$\ln\gamma_i^{R\acute{e}s} = v_{k,i}\left(\ln\Gamma_k - \ln\Gamma_{k,i}\right)$$

Where Γ_k is the residual activity coefficient of group k and $\Gamma_{k,i}$ is the residual activity coefficient of group k, in a reference solution containing only molecules of type i.

The natural logarithm of the activity coefficient of the group k, whether in the mixture or in the pure body i is given by :

$$\ln\Gamma_k = Q_k\left[1 - \ln\sum_m \theta_m^g \Psi_{mk}\right] - \sum_m \frac{\theta_m^g \Psi_{mk}}{\sum_n \theta_n^g \Psi_{nm}}$$

The area fraction of group m is calculated in the same way as θ_i :

$$\theta_m^g = X_m^g Q_m \Big/ \sum_n Q_n X_n^g$$

Where X_m^g is the molar fraction of group m in the mixture.

$$X_m^g = \sum_j x_j v_{jk} \Big/ \sum_j x_j v_j$$

Ψ_{nm}: the interaction parameter between groups n and m.

$$\Psi_{nm} = \exp\left(-(U_{nm} - U_{mm}/RT)\right) = \exp(-a_{nm}/T)$$

Where U_{nm} is the interaction energy between the n and m groups.

a_{nm} interaction parameter evaluated from the experimental results.

With : $a_{m,n} \neq a_{n,m}$

b) UNIFAC modified

The original UNIFAC model has shortcomings with respect to proximity effects, or mixtures containing isomers, or good estimation of activity coefficients at infinite dilution, or liquid-liquid equilibria or enthalpies of mixtures. To overcome these shortcomings, due mainly to the interaction parameters being independent of temperature, various modifications have been proposed. They are mainly of two types:

- Modification of the combinatorial part and in particular the volume fraction Qi, in order to better estimate the activity coefficients at infinite dilution.
- Consideration of the variation of binary interaction parameters between am,n groups as a function of temperature.

✓ *Modified UNIFAC (Lyngby version)*

The combinatorial term is given by the following expression:

$$\ln\gamma_i^{comb} = \ln\frac{w_i}{x_i} + 1 - \frac{w_i}{x_i}$$

$$w_i = \frac{r_i^p x_i}{\sum_j r_j^p x_j}$$

The exponent p is relative to the volume parameter r_i of the molecule; it has been fixed at the value of 2/3.

$$r_i = \sum_k v_k^i R_k$$

The residual part is given by the following expression :

$$\ln\gamma_i^{Rés} = \sum \ v_k^i\left(\ln\Gamma_k - \ln\Gamma_k^i\right)$$

$$\ln\Gamma_k = q_k\left[1 - \ln\left(\sum_{m=1}^{NG}\Theta_m\Psi_{m,k}\right) - \sum_{m=1}^{NG}\left(\Theta_m\Psi_{k,m}\right)\Big/\left(\sum_n\Theta_n\Psi_{n,m}\right)\right]$$

The interaction parameter depends on temperature and is expressed as follows:

$$\Psi_{n,m} = \exp\left[-\frac{A_{n,m}}{T}\right]$$

$$\text{With :} A_{n,m} = a_{n,m} + b_{n,m}(T - T_0) + c_{n,m}\left(T\ln\left(\frac{T}{T_0}\right) + T - T_0\right)$$

Where T0 is the reference temperature, taken arbitrarily equal to 298.15 K.

$a_{n,m}$; $b_{n,m}$ and $c_{n,m}$ are the interaction parameters.

✓ *Modified UNIFAC (Gmehling version)*

In this version, the term combinatorial is described by the expression below:

$$\ln\gamma_i^{comb} = 1 - \phi_i + \ln\phi_i - 5q_i[1 - (\phi_i/\theta_i) + \ln(\phi_i/\theta_i)]$$

With $\quad \theta_i = r_i^{3/4}/\sum_j x_j r_j^{3/4} \quad$ and $\quad \phi_i = r_i/\sum_j x_j r_j$

On the other hand, the group activity expression is the same as that described in the original version. The dependence of the interaction parameters on temperature is given by the following expression:

$$\Psi_{nm} = \exp\left(-a_{nm}/T\right)$$

$$a_{nm} = a1_{nm} + a2_{nm}T + a3_{nm}T^2$$

III.2.2. Chemical interaction models

Physical interaction models describe the non-ideality of solutions in terms of non-specific physical intermolecular forces. The activity   coefficients estimated by these models are related to physical parameters (size of the molecules). Nevertheless, deviations from ideality can still exist despite their estimation through these models. These deviations result from chemical reactions which are of two types: association and solvation reactions.

III.2.2.1. Solvation models

The thermodynamic study of the solvation of a solute-solvent couple implies the decomposition of the total free energy involved, in terms of contributions corresponding to the interactions of different natures.

This approach has led to the development of Linear Free Energy Relationships (LFER), including chemical or physico-chemical parameters, capable of quantifying the variations of the free energy during the changes of state of a system.

III.2.2.1.1. Model of Kamlet, Taft et al

The LFER concept developed by Kamlet, Taft et al has been used to explain the effects of solvent on the spectroscopic properties of solutes, reaction rates, chemical equilibria, and retention phenomena in chromatography.

The parameters introduced in their models are called "solvatochromic" parameters and are determined from the known physical properties of the substance to arrive at the solvation energy.

This approach allowed the prediction of solubility and partition coefficient.

Solvation of a solute molecule consists of three steps:

> ➤ Formation of the cavity in the solvent of volume equal to that of the molecule of the

solute. The corresponding free energy is $(\Delta G^{solv})_{cav}$ It is given by the following expression :

$$(\Delta G^{solv})_{cav} = A\delta_1^2 \frac{V_2}{100}$$

Where V_2 is the molar volume of the solute, δ_1 is the Hildebrand solubility parameter and A is a constant.

> ➤ Reorganization of the solvent molecules around the cavity and reorganization

conformation of the solute inside the cavity. The corresponding free energy is $(\Delta G^{solv})_{réorg}$. This term is not considered, because of its small value.

> ➤ Keesom and Debye interaction between the solute molecule and the neighbouring solvent molecules $(\Delta G^{solv})_{DP}$ and the formation of a hydrogen bond corresponding to $(\Delta G^{solv})_{HB}$.

$(\Delta G^{solv})_{DP}$ can be calculated by the expression :

$$(\Delta G^{solv})_{DP} = B.\,\pi_1^*.\,\pi_2^*$$

Where π_1^* *et* π_2^* are the solvatochromic dipolarity/polarizability parameters of the solvent molecules and the solute molecule respectively. B is a constant.

The formation of a hydrogen bond is accompanied by a Gibbs energy change, given by :

$$(\Delta G^{solv})_{HB} = C\alpha_1\beta_2 + D\alpha_2\beta_1$$

Where α_1 and α_2 represent the ability of the solvent and the solute to form a hydrogen bond, respectively, β_1 and β_2 are respectively the parameters expressing the ability of the solvent and the solute to accept hydrogen bonding.

According to Taft et al [19,20], the free energy of solvation is equal to:

$$(\Delta G^{solv})_{total} = (\Delta G^{solv})_{cav} + (\Delta G^{solv})_{réorg} + (\Delta G^{solv})_{DP} + (\Delta G^{solv})_{HB}$$

The LFER model of Taft et al is then written :

$$(\Delta G^{solv})_{total} = A\delta_1^2 \frac{V_2}{100} + B.\pi_1^*.\pi_2^* + C\alpha_1\beta_2 + D\alpha_2\beta_1$$

III.2.2.1.2. LSER model

In order to quantify the intermolecular solute-solvent interactions, Abraham and his collaborators developed the solvation model, allowing the correlation of thermodynamic properties.

The general solvation equation used by Abraham et al is given by :

$$\mathbf{logSP = c + eE + sS + aA + bB + lL}$$

Where SP is a solvation parameter. SP can be the net retention volume per gram of impregnated support of the stationary phase V_N^* or the gas-liquid partition coefficient of the solute K_L in the stationary phase.

E represents the molar excess refraction, it was defined as the molar refraction of the solute minus the molar refraction of an n-alkane having the same McGowan volume:

$$\mathbf{E = MR(soluté) - MR(alcane\ ayant\ le\ même\ volume\ V_x)}$$

Where MR is the molar refraction defined by the Lorentz-Lorenz relation:

$$\mathbf{MR = V_x \frac{\eta^2-1}{\eta^2+2}}$$

η is the refractive index of a liquid at 20°C.

V_x is the McGowan volume, whose unit is Cm3mol-1/10.

S characterizes the polarizability of the molecules

The parameter L, which corresponds to the decimal logarithm of the Ostwald solubility coefficient for a solute in n-hexadecane at 25°C L16

$$\log L^{16} = \frac{\textbf{concentration du soluté dans le n − hexadécane}}{\textbf{concentration du soluté dans la phase gazeuse}}$$

This parameter is measurable by IGC and corresponds to the gas-liquid partition coefficient of a solute injected into a column containing hexadecane as stationary phase.

L16 is used to characterize the Gibbs energy of cavity formation and solute-solvent dispersion interactions.

The Abraham model based on experimentally measurable descriptors is used to determine the LSER parameters of the c, e, s, a, b and l ionic liquids.

e represents the contribution of the ionic liquid interactions with the π-bonds or n-electrons of the solute.

s represents the contribution of dipole-dipole or induced dipole-dipole interactions.

a characterizes the basicity of the ionic liquid.

b characterizes the acidity of the ionic liquid.

l represents the ability of the ionic liquid to separate a series of compounds from the same family.

The determination of the LSER parameters of ionic liquids is done by injecting n reference probes to obtain n equations of the type :

$$\log K_L = c + eE + sS + aA + bB + lL$$

The constants c, e, s, a, b and l are determined by multiple linear regression.

III.2.2.2. Models of the associated solutions

Definition:

We have seen systems whose heat of mixing was :

- Either negligible: athermal solution

- Either of the same order as the thermal agitation energy: regular solution

Let us now consider classes of solutions in which the heat of mixing reaches much higher values (of the order of 5000 cal/mol).

Example: solution containing acetic acid in a dimeric form (establishment of the hydrogen bond).

The interaction between the two molecules forming the dimer will therefore have changed the rotational and vibrational state of each.

Generally speaking, we can classify the molecules in two categories:

1. Those whose rotational or vibrational state is not modified: monomolecules.
2. Those whose state of vibration or rotation is, on the contrary, modified by the presence of neighbouring molecules. These molecules are associated and form associated complexes.

Thermodynamic properties of associated solutions

Let there be a solution (A+B), let us assume that there exist in this solution complexes Aγ formed by γ monomolecules A, complexes Bγ and finally complexes AiBj, resulting from the association of i molecules of A and j molecules of B.

Let nAγ, NBγ, nAiBj be the number of moles of these complexes.

We'll get:

$$n_A = \sum_{\gamma} \gamma n_{A\gamma} + \sum_{i} \sum_{j} i n_{A_i B_j}$$

$$n_B = \sum_{\gamma} \gamma n_{B\gamma} + \sum_{i} \sum_{j} j n_{A_i B_j}$$

In the case where A alone is associated, the previous equations reduce to:

$$n_A = \sum_{\gamma} \gamma n_{A\gamma}$$

Similarly, if next to the monomolecules, there are only AB complexes, we will have:

$$n_A = n_{A_1} + n_{AB}$$

$$n_B = n_{B_1} + n_{AB}$$

Let us denote by: $\begin{Bmatrix} \mu_{A\gamma} \\ \mu_{B\gamma} \\ \mu_{A_iB_j} \end{Bmatrix}$ Chemical potentials of different complexes in solution.

$\begin{Bmatrix} \mu_A \\ \mu_B \end{Bmatrix}$ Macroscopic chemical potentials of components A and B.

Let us propose to find a link between the macroscopic chemical potentials and the chemical potentials of complexes.

Let us first express that the different complexes are in equilibrium with each other, as well as with the monomolecules A1 and B1. Let us consider the reactions:

$$A_\gamma \leftrightarrow \gamma A_1$$

$$B_\gamma \leftrightarrow \gamma AB_1$$

$$A_iB_j \leftrightarrow i_{A_1} + j_{B_1}$$

The chemical potentials are given by the expressions :

$$\begin{Bmatrix} \mu_{A\gamma} = \gamma\mu_{A_1} \\ \mu_{B\gamma} = \gamma\mu_B \\ \mu_{A_iB_j} = i\mu_{A_1} + j\mu_{B_1} \end{Bmatrix}$$

At constant T and P, let us write the total differential of G in the associated system, considered as a mixture of different complexes, we will have taking into account the above systems of equations :

$$dG = \sum \mu_{A_\gamma} dn_{A_\gamma} + \sum \mu_{B_\gamma} dn_{B_\gamma} + \sum_i \sum_j \mu_{A_iB_j} dn_{A_iB_j} = \sum \gamma \mu_{A_1} dn_{A_\gamma} + \sum \gamma \mu_{B_1} dn_{B_\gamma} + \sum_i \sum_j (i\mu_{A_1} + j\mu_{B_1}) dn_{A_iB_j}$$

$$dG = \mu_{A_1} dn_A + \mu_{B_1} dn_B$$

At constant T and P for a binary mixture :

$$dG = \mu_A dn_A + \mu_B dn_B$$

We can conclude from the previous equations that :

$$\mu_A = \mu_{A_1}$$

$$\mu_B = \mu_{B_1}$$

The macroscopic chemical potentials are therefore equal to the chemical potentials of monomolecules.

Let us consider the solution formed by the monomolecules and the complexes and note that between these different molecular species any interaction that could lead to an association is excluded by definition. It follows that the monomolecule-complex system must approach an ideal system.

By neglecting the deviations from ideality which could appear because of the differences in dimensions and shapes of the particles, we are thus led to the model of an associated ideal solution. In such a solution, the chemical potentials of the monomolecules are of the form :

$$\mu_{A_1} = \mu_{A_1}^*(T, P) + RT \ln N_{A_1}$$

$$\mu_{B_1} = \mu_{B_1}^*(T, P) + RT \ln N_{B_1}$$

Where N_{A_1} and N_{B_1} are the effective titers of the monomolecules A and B, i.e. :

$$N_{A_1} = \frac{n_{A_1}}{\sum n_{A_\gamma} + \sum n_{B_\gamma} + \sum_i \sum_j n_{A_i B_j}}$$

$$\gamma_A = \frac{N_{A_1}}{N_A} \exp \frac{\mu_{A_1}^* - \mu_A}{RT}$$

$$\gamma_B = \frac{N_{B_1}}{N_B} \exp \frac{\mu_{B_1}^* - \mu_B}{RT}$$

$$N_{B_1} = \frac{n_{B_1}}{\sum n_{A_\gamma} + \sum n_{B_\gamma} + \sum_i \sum_j n_{A_i B_j}}$$

So:

$$\mu_A^*(T, P) + RT \ln N_A f_A = \mu_{A_1}^*(T, P) + RT \ln N_{A_1}$$

$$\mu_B^*(T, P) + RT \ln N_B f_B = \mu_{B_1}^*(T, P) + RT \ln N_{B_1}$$

Hence:

In the symmetric reference system :

$$N_A \to 1, \gamma_A \to 1$$

Case where only the molecules of one of the two components, i.e. A, are associated with each other

It is more advantageous to use a very dilute solution of A in B as a reference system.

Indeed, if $N_A \to 0, \gamma_A \to 1$

The result is that: $\gamma_A = \dfrac{N_{A_1}}{N_A}$

On the other hand, when $N_B \to 1, \gamma_A \to 1$ and $N_{B_1} \to 1$, then :

$$\gamma_B = \frac{N_{B_1}}{N_B}$$

Application exercises

<u>**Exercise 1**</u>

The boiling temperatures of six mixtures of pentane (1) and hexane (2) as a function of the composition of the liquid and vapour phases are given in the table below:

Teb (°C)	68,2	63,1	54,5	47,5	41,2	35,6
x1	0	0,108	0,310	0,522	0,747	1
y1	0	0,231	0,539	0,750	0,897	1

The binary is treated as a mixture.

1. Calculate the partial pressure of pentane P1 in each mixture. What is the partial pressure that pentane would have if the mixture were ideal? Calculate the relative difference between the two values : $\frac{P_1 - P_1^{id}}{P_1^{id}}$ (in %).

2. The mixture can be represented by the regular solution equations :

   $$ln\gamma_1 = ax_2^2 \text{ and } ln\gamma_2 = ax_1^2$$

   Calculate $ln\gamma_1$ for each mixture. Deduce the average value of the parameter a, assumed to be independent of temperature.

3. Calculate the partial pressure Pireg given by the model. What is the relative difference between P1 and Pireg : $\frac{P_1 - P_1^{reg}}{P_1^{id}}$ (in %).

The saturation vapour pressure of pentane is given:

$$log_{10}P_1^*(\text{mmHg}) = 6,87715 - \frac{1075,78}{T_{eb} + 233,205}$$

<u>**Exercise 2**</u>

The molar free enthalpy of excess of the methanol (1) - trichloroethylene (2) mixture depends on the molar titers according to the characteristic reaction of strictly regular mixtures:

$$G^E = wx_1x_2$$

1. Establish the relationships giving the chemical potentials of excess methanol and trichloroethylene as a function of the composition of the mixture.
2. Establish the expression for the molar enthalpy of excess ^{HE} as a function of composition. Also give the expression for the molar enthalpy of mixing ^{HM}.
3. What is the value of the excess molar entropy?
4. What can be said about the molar entropy of mixing?
5. Give the expression for the molar free enthalpy of mixing ^{GM}.
6. The methanol-trichloroethylene mixture is an azeotrope. Show that at the azeotropic composition of a mixture at a given temperature, the following relationship relates the vapour pressures of the pure substances to the activity coefficients in the liquid phase:

$$[\gamma_1 P_1^* = \delta_2 P_2^*]_{az}$$

7. Calculate the composition of the azeotrope and the equilibrium pressure $P_{éq}$ at 330 K and 350 K.

Data: w=5391 J.mol-1

Saturation vapour pressures of methanol and trichloroethylene:

$$\ln P_1^*(\mathbf{mmHg}) = 20,741 - \frac{4765,5765}{T}$$

$$\ln P_2^*(\mathbf{mmHg}) = 17,71426 - \frac{3986,9220}{T}$$

Exercise 3

In a liquid binary mixture, the rational activity coefficient of component 1 is related to the mole fraction of component 2 by the relation :

$$\ln \gamma_1 = \frac{w}{RT} x_2^2$$

W is a constant independent of temperature.

1. From the Gibbs-Duhem equation, find the relationship between the activity coefficient γ_2 to the mole fraction x1.

2. The liquid-vapour equilibrium of the system is studied. The activity coefficient γ_1 of component 1 in the liquid phase obeys the above equation. P1 and P2 are the partial pressures of the two components in the vapour phase, assumed to be perfect, in equilibrium with the liquid at temperature T P_1^* and P_2^* are the saturation vapour pressures of 1 and 2 at the same temperature. Write the relations linking P1 and P2 to the molar fractions x1 and x2 of the liquid, the activity coefficients γ_1 and γ_2 and the saturation vapour pressures P_1^* and P_2^*.

3. The partial molar free enthalpy of excess g_1^E of silver chloride in the molten salt mixture AgCl-LiCl is related to the mole fraction x1 by the following law at 700°C.

x1	0,1	0,3	0,5	0,7	0,9
g_1^E (J.mol-1)	7110	4301	2195	790	88

Show that this mixture obeys the relation (1) and give the value of w. Calculate

γ_1 and γ_2 for the various compositions of the painting.

Correction of exercises

CHAPTER I

Exercise 1

1. For a component in an ideal liquid mixture : $\mu_i(T, P, l) = \mu_i^* T, P, l + RTLnx_i$

2. a. The molar masses of water and methanol are: $M_{(H2O)}$=18 g/mol and M(CH3OH)=32 g/mol. The molar numbers of water and methanol are therefore respectively: $n_{(H2O)}$=50 mol and n(CH3OH)=2.5 mol. Hence the molar fractions : $x_{H2O}^l = 0.95 \ et \ x_{CH3OH}^l = 0.48$

According to Raoult's law applicable in the case of ideal mixtures, the partial pressures of the vapors that overcome the liquid mixture are : $P_{H2O} = x^l_{H2O}P^*_{H2O} = 17.1\ mmHg$ and $P_{CH3OH} = x^l_{CH3OH}P^*_{CH3OH} = 4.5\ mmHg$

The total steam pressure is therefore : $P_T = P_{H2O} + P_{CH3OH}$=21.6 mmHg.

    b. The partial pressures are related to the total pressure by the expression :$P_i = x^v_i P_T$

Therefore: xvH2O=0.792 and xvCH3OH=0.208

The vapour is therefore richer in methanol than the liquid, methanol being the most volatile component.

Exercice 2

1. The total pressure is equal to the sum of the partial pressures: P=P1+P2

   From the diagram, if Pi* is the saturation vapour pressure of component i :

   $$P_2 = P^*_2 x(CCl4) \text{ and } P_1 = P^*_1 x(SnCl4)= P^*_1(1 - x(CCl4))$$

2. Each of the constituents of the mixture thus verifies Raoult's law.

   For each component i, the equilibrium between the liquid and its vapour results in the equality of the chemical potentials: $\mu_i(T,P,v)=\mu_i(T,P,l)$.

   For $_{Bi}$ gas, the vapors are assumed to be perfect: $\mu_i(T,P,v) = \mu^0_i(T,v) + RTln(\frac{P_i}{p^0})$

   For B, liquid, using the experimental results studied in the first question :

   $$\mu_i(T,P,l) = \mu_i(T,P,v) = \mu^0_i(T,v) + RTln(\frac{P_i}{p^0})= \mu^0_i(T,v) +$$

   $$RTln\frac{(P^*_i)}{p^0}+RTlnxi$$

   If the compound Ai is pure under standard conditions, we have

   $$\mu^0_i(T,l) = \mu^0_i(T,v) + RTln\frac{(P^*_i)}{p^0} \text{ so} : \mu_i(T,P,l) = \mu^0_i(T,l)+RTlnxi$$

<u>**Exercise 3**</u>

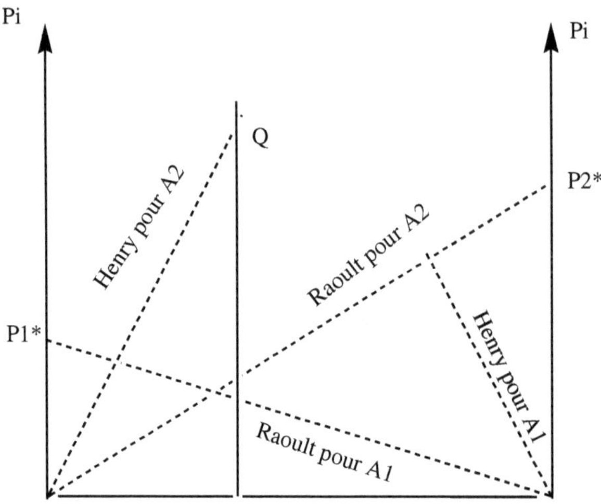

b. In mixed convention : $\mu_i(T, l) = \mu_i^0(T, l) + RTLna_i$

Assuming perfect vapors : $\mu_i(T, v) = \mu_i^0(T, v) + RTln(\frac{P_i}{p^0})$

Writing the equality of chemical potentials at equilibrium, we obtain:

$$\mu_i^0 T, l) + RTLna_i) = \mu_i^0(T, v) + RTln(\frac{P_i}{p^0})$$

The activity coefficient γi has the expression: $\gamma_i = \frac{a_i}{x_i}$

$$\frac{P_i}{a_i P^0} = \frac{P_i}{\gamma_i x_i P^0} = exp\left[\frac{\mu_i^0(T, l) - \mu_i^0(T, v)}{RT}\right]$$

If we approach the pure constituent :

$$\lim_{x_i \to 1} \frac{P_i}{\gamma_i x_i P^0} = \frac{P_i^*}{P^0} =$$

After identification, $P_i = \gamma_i x_i P_i^*$ and we finally obtain : $\gamma_i = P_i / P_i^* x_i$

c. If we refer to the infinitely diluted state of the solution, the expression of the chemical potential is: : $\mu_i(T, l) = \mu_{i,x,\infty}^0(T, l) + RTLna_i$

$$\lim_{x_i \to 1} \frac{P_i}{\gamma_i x_i P^0} = \frac{ki}{P^0} \text{ we access } \gamma_i = \frac{P_i}{x_i ki}$$

__Exercice 4__

a. The solutions being ideal. The reference state for each is the pure liquid body at T under its saturation vapour pressure, i.e. $a_1 = \frac{P_1}{P_1^{pur}}$, $a_2 = \frac{P_2}{P_2^{pur}}$.

Moreover, a1=x1, a2=x2. We deduce :

$$P_1 = x_1 P_1^{pur}, P_2 = x_2 P_2^{pur} \text{ et } P_T = P_1 + P_2 = x_1(P_1^{pur} - P_2^{pur}) + P_2^{pur}$$

Hence the graph:

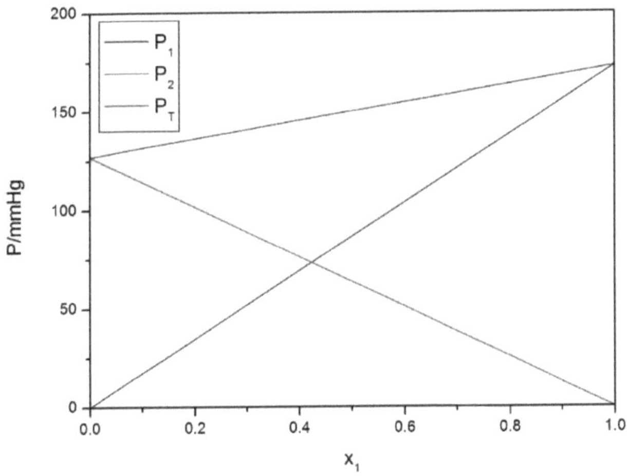

b). M(1)=180 g/mol, M(2)=202 g/mol

so x1=0.1184

PT=0.1184(173-127)+127=132.45 mmHg

c). We apply Dalton's law to the gas phase, i.e. $y_1 = \frac{P_1}{P_T}$

Then y1=0.1546 and y2=1-y1=0.8454

(d) If $n_1^{vap} = n_2^v, y_1 = y_2$ et $P_1 = P_2$

$$x_1 P_1^{pur} = (1 - x_1) P_2^{pur}$$

Hence x1=0.4233 and x2=0.5767

Exercise 5

The solution (1)-(2) being ideal, we have

$$a_1 = x_1 = \frac{P_1}{P_1^{pur}}$$

a1 is the activity of (1) in the liquid solution, the reference state being pure liquid (1) at 137°C. x1 is the molar titre (composition) of (1) in the liquid phase. P1 is the partial pressure of (1) in the gas phase in equilibrium with the solution at 137°C.

P1 is the saturation vapour pressure of (1) at 137 °C.

Also:

$$a_2 = x_2 = \frac{P_2}{P_2^p} \quad \text{either,} \quad P_2 = x_2 P_2^{pur} = (1 - x_1) P_2^{pur}$$

We deduce, $P_T = P_1 + P_2 = x_1(P_1^{pur} - P_2^{pur}) + P_2^{pur}$

x1=0P1=0P2=P2pur=PT

x1=1P1=P1pur=PTP2=0

2) a. From $P_T = x_1(P_1^p - P_2^p) + P_2^p$ we derive :

$$x_1 = \frac{P_T - P_2^p}{P_1^p - P_2^p} = 0.7487$$

It is of course possible to read this result directly from the graph.

b. From Dalton's Law: $y_1 = \frac{P_1}{P_1^p} = 0.85$

c. $n_1^l = n_1^v$ entraine $x_1^l = x_1^v = 0.5$

and PT=0.5(863-453)+453=658 mmHg

We deduce y1=0.6558.

CHAPTER II

<u>Exercice 1</u>

Representation of PT(X2), and Pi (X2), where X2 is the mole fraction of propanol in the liquid phase.

PT=P1+P2. According to Dalton's law: Pi=(xi)v*PT

The values found for PT are summarized in the table below:

(X2)L	0	0.02	0.05	0.1	0.2	0.4	0.6	0.8	0.9	0.95	1
P2 mmHg	0	5.05	10.8	12.7	13.6	14.2	15.5	17.8	19.4	20.8	21.8
P1 mmHg	23.8	23.5	23.2	22.7	21.8	21.7	19.9	13.4	8.1	4.2	0.0
PT mmHg	23.8	28.55	34.0	35.4	35.4	35.9	35.4	31.2	27.5	25	21.8
(x2)v	0	0.177	0.317	0.359	0.384	0.395	0.438	0.570	0.705	0.832	1

We obtain the isothermal diagram at pressure maxima. The system presents a negative deviation, which is translated by the presence of an azeotrope at maxima.

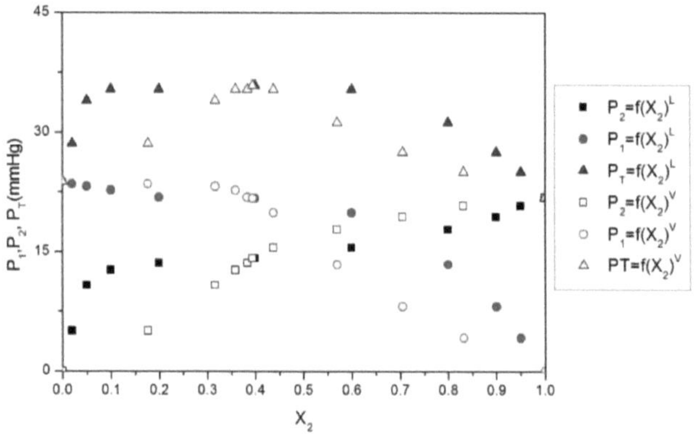

Coordinates of the azeotrope: Xaz=0.4; Pmax=35.9 mmHg.

The plot of the curves Pi=f(Xi)L shows a positive deviation.

2- a)- Discussion of the curves :

In the vicinity of (X2)L =0 (solute: propanol; solvent: water)

For A2 we apply Henry's law: $P_2 = k_{H,2}(X_2)_L \rightarrow k_{H,2} = \dfrac{5.05}{0.02} = 252.5\ mmHg$

For A1 (water) we apply Raoult's law: $P_1 = P_1^* X_1^L$

In the vicinity of (x2)L=1 (solute: water, solvent: propanol)

For A2 Raoult's Law $P_2 = P_2^* X_2^L$

For   A1 Henry's Law $P_1 = k_{H,1}(X_1)_L$ with $k_{H,1} = \dfrac{4.2}{(1-0.95)} = 84\ mmHg$

b)- Activity coefficients γi knowing that the reference chemical potentials correspond to pure bodies.

$$a_i = \gamma_i X_i^L = \frac{P_i}{P_i^*}$$

For (X2)L=0.2 γ2=3.12 and γ1=1.145

c) - Evaluation of the activity coefficients at infinite dilution

$$a_i^\infty = \gamma_i^\infty X_i^L = \frac{P_i}{k_{H,i}}$$

So for $(x2)_{L=0.2}$ $\gamma_2^\infty = 0.2693$ and $\gamma_1^\infty = 0.3244$

3)- Composition of the condensed vapour which is determined graphically by extrapolation on the dew curve and we obtain $(X2)_{V=0}.83$ in propanol.

Exercise 2

a)- The isobaric diagram gives T as a function of the mass percentage (mass fraction WHNO3 from 0 to 100%).

V: single-phase steam system

L: single-phase liquid system

L+V: two-phase liquid+steam system

R: dew point curve

E : boiling curve

The liquid mixture whose composition corresponds to the abscissa of the maximum is an azeotrope (at minimum pressure: Preelle=P1+P2<Pideal.

Property : by boiling the azeotropic mixture, we obtain a vapour of the same composition xv=xL=xaz

b) - calculate the masses of each component:

For HNO3: m1=n1*M1=18.9 g

For H2O: m2=n2*M2=(n-n1)M2=66.6 g

The total mass of the mixture: m=m1+m2=85.5 g

The molar fraction of HNO3: WHNO3=(m1/m)*100=22.10

At tI the figure point belongs to the liquid range (tI=100°C; WI=22.10)

- The beginning of boiling is for t1 given by the boiling curve (E) which is determined graphically and we find t1=105.5°C.

c)- By application of the chemical moment rule :

For a heating, leading to M' in the two-phase system, we have 2 phases in equilibrium:

1 vapour phase of composition $(WHNO3)_{V=0.10}$

1 liquid phase of composition $(WHNO3)_{L=0.40}$

Chemical moment rule :

$mV/mL=(0.4-0.22)/(0.22-0.1)$.

Exercise 3

1) A diagram with a eutectic point E is obtained.
   - Reciprocal zero solubility in the solid state, thus 2 distinct solid phases (solid α-naphthol and solid naphthalene)
   - Total solubility in the liquid state.
1) A) Coordinates of M0 initial liquid mixture :

$$n\alpha\text{-naphthol}=40/144 =(n2)_0$$

$$n\text{naphtalène}=10/128=(n1)_0$$

$$X0\text{naphtalène}= (n1)0/(n1)0+(n2)0=0.22$$

So the coordinates of point M0 are : $(X1)=0.22; T0=100°C$

The first crystals form when the liquidus curve is reached:

Graphically: $T1=87.5°C$

Nature of crystals: pure α-naphthol

b)- Construction and reading of the thermal diagram :

M0M1→A0A1 Liquid cooling from $T0=100°C$ to$T1=87.5°C$

M1→A1 appearance of the first crystals of pure α-naphthol

M1E→A1AE crystallization of α-naphthol and enrichment of the liquid phase with naphthalene (figurative point for liquid moving on liquidus

ME→E First appearance of pure naphthalene crystals.

AEA'E Simultaneous crystallization of α-naphthol and naphthalene, giving separate solid phases (T_E=61°C=Constant as long as liquid remains) In A'E : disappearance of the last drop of liquid

MEM4→A'$_{4}$A4 cooling of the two solid phases.

c) - Application of the chemical moment rule for two temperatures:

> Mixture cooled down to 75°C : point M_2 (two-phase system)
>> A liquid phase of composition X_{a2}=0.415 in naphthalene and 0.585 in α-naphthol
>> A solid phase of pure α-naphthol X_{b2}=0 in naphthalene.
>> We have: $nL(0.415-0.22)=nS(0.22)$ knowing that nL and nS are respectively the number of moles of liquid and solid (α-naphthol)
>> Now $nL+ns=(n1)_0+(n2)_0=0.356$ mol and therefore nL=0.356-nS
>> So ns=n α-naphthol=0.1673 mol so m α-naphthol solid=24.01g
>> And nL=n1+n2=n-ns=0.1887 mol.
>> Hence nnaphthalene=n1=X_{a2}*nL=0.415*nL=0.0783 mol.
>> And nα-naphthol=n2=(1-X)nL=0.1104 mol
>> Hence in mass: mnaphthalene=10.022 g(liquid)
>> mα-naphthol=15.9 g(liquid)

> Mixture cooled to 50°C

This temperature is lower than T_E, so the figure point is in the two-phase system of the two solid phases.

a solid phase of pure α-naphthol of mass =40 g

a solid phase of pure naphthalene of mass =10g

Exercise 4

1) Iron-cementite
2) Peritectic reaction

3) Eutectoid reaction

4) Eutectoid steel

     (a) Austenite γ

     b) 1400°C

5 - Nature of the phases: liquid, L+α, ...

6- At 1147°C the solubility of carbon is maximum in the γ phase.

7- Maximum percentage of carbon in ferrite α is 0.02.

8- A 722-ε°C, the proportions of the phases formed by a 1% carbon alloy are :

    m (α phase)=85.47%; m(Fe3C cementite)=14.53%.

Exercise 5

1- Nature of phases :

1: a homogeneous liquid phase; 2: 1 liquid phase and a solid α phase; 3: 1 solid α phase; 4: 1 liquid phase + 1 solid β phase; 5: 1 solid β phase; 6: 1 solid α phase + 1 solid β phase.

solid α : AgNO3 in solid solution in NaNO3

solid β : NaNO3 in solid solution in AgNO3

2. a- Consider cooling the liquid to 22 mol% NaNO3 :

T>222°C: a homogeneous liquid phase with 22% mol NaNO3 is present.

218°C<T<222°C: Deposition of solid α whose composition varies from about 41% (first crystal) to 38% NaNO3. The liquid decreasing from 22% to 20% NaNO3.

At T=218°C: a first crystal of solid β of composition 24% NaNO3 appears. Three phases are present; a binary peritectic reaction at constant T and compositions will occur.

Liq (20%NaNO3)+Sol α (38% NaNO3)→Sol β (24%NaNO3)

215°C<T<218°C: there is deposition of solid β, the last drop of liquid disappears.

207.5°C<T<215°C: a solid phase β at 22% NaNO3.

T<207.5°C: two solid phases α whose composition increases from 40% NaNO3 and β, whose composition decreases from 22% NaNO3.

b. Consider cooling the liquid to 30%mol NaNO3 :

T>236°C: a homogeneous liquid phase with 30% mol NaNO3 is present.

218°C<T<236°C: there is deposition of solid α whose composition varies from about 49% (first crystal) to 38% NaNO3, the liquid goes from 30 to 20% NaNO3.

At T=218°C: a first crystal of solid β of composition 24% NaNO3 appears. Three phases are present. A binary peritectic reaction at constant T and compositions will occur, until the last drop of liquid disappears.

Liq (20%NaNO3)+Sol α (38% NaNO3)→Sol β (24%NaNO3)

T<218°C: two solid phases remain, β whose composition decreases from 24% NaNO3 and α whose composition increases from 38% NaNO3.

c. Consider the cooling of the liquid to 60% NaNO3 :

T>275°C: a homogeneous liquid phase at 60% NaNO3 is present.

275°C>T>256°C: there is deposition of solid α whose composition varies from about 72% NaNO3 (first crystal) to 60% NaNO3. The liquid changes from a composition of 60% to 43% NaNO3 (last drop).

T<256°C: only a solid phase α to 60% NaNO3 remains.

<u>Exercise 6</u>

1. For the domains scored from 1 to 9, we have :

1 : 1 liquid; 2 : 1 liquid + CaO solid; 3 : 1 liquid + CaO,TiO2 solid; 4 : 1 liquid + CaO,TiO2 solid; 5 : 1 liquid +3CaO,2TiO2 solid; 6 : CaO solid + 3CaO,2TiO2 solid; 7 : CaO,TiO2 solid + 3CaO,2TiO2 solid; 8 : 1 liquid + TiO2 solid

9: CaO, solid TiO2 + solid TiO2

   2. a. Cooling of 45% TiO2 liquid

T> 1800°C : 1 homogeneous liquid at 45% TiO2

1750°C<T<1800°C: appearance of the solid phase CaO,TiO2. The liquid changes from the composition of 45% to 40.5%.

T=1750°C : binary peritectic reaction :

Liq (40.5%TiO2) + solid CaO,TiO2 $\rightarrow$ solid 3CaO,2TiO2

Until disappearance of solid CaO,TiO2

1750°C>T>1696°C: 1 liquid phase from 40.5 to 37% TiO2 and a solid phase 3CaO,2TiO2.

T=1696°C : binary eutectic reaction :

Liquid (37% TiO2)$\rightarrow$CaO solid+3CaO,TiO2 solid

Until the liquid disappears.

b. Cooling of the liquid to 55% TiO2 :

Binary peritectic reaction: **Liq (40.5%TiO2) + CaO,TiO2 solid $\rightarrow$ 3CaO,2TiO2 solid**

Until the liquid disappears.

c. Cooling of the liquid to 70% TiO2 :

Binary eutectic reaction: **liq (70%TiO2)$\rightarrow$ CaO,TiO2 solid + TiO2 solid**

Until the liquid disappears.

3. At 1800°C, the initial mixture is 45% TiO2 and is in two phases:

- A 43% TiO2 liquid
- A CaO,TiO2 solid.

With a mass ratio: liq/sol=14/2=7

At 1733°C, the initial 45% TiO2 mixture is in two phases:

- A 39% TiO2 liquid; A 3CaO,2TiO2 solid with a mass ratio liq/sol=4/6=0.66

At 1600°C: the initial 45% TiO2 mixture is in two phases:

- A CaO solid; A 3CaO,2TiO2 solid with a mass ratio CaO/3CaO,2TiO2=4/45=0.09.

Exercise 7

1- The different phases in the different areas:

1: Liquid; 2: Liq + A; 3: A+D; 4: Liq +D; 5: Liq + B; 6: D+B

2- Liquid cooling to 70% of B :

Binary peritectic reaction at 800°C: **liq (60%B) + solid B → solid D**

Until the disappearance of B.

Binary eutectic reaction at 700°C: **Liq (35% B) → A solid + D solid**

Until the liquid disappears.

3- Compound D is 80% of B or AB4.

Below is the lead-bismuth (Pb-Bi) binary equilibrium diagram

4. Peritectic reaction with a composition of 35.8% Bi at T=184°C.

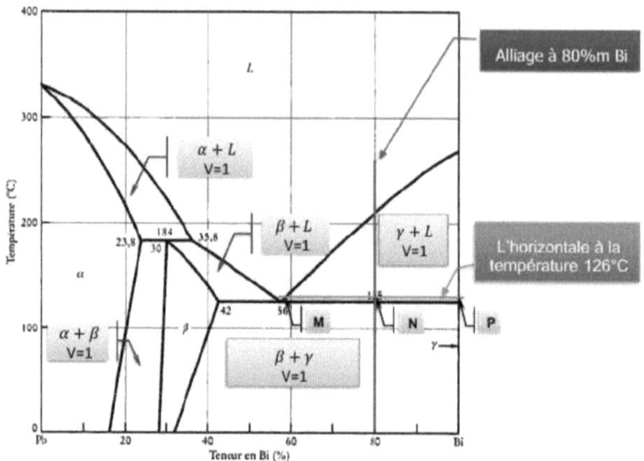

Eutectic reaction with a composition of 56% Bi and a temperature of T=125°C.

The phases present at T=126°C are the liquid phase with a composition of 56% Bi and the solid phase γ with a composition of 99.9% Bi.

5. Estimation of the mass fractions of the phases in equilibrium present at T=126°C, for an alloy of 80%-mass Bi.

Applying the chemical moment theorem we obtain: $x_{L=0}$**.45; Xγ=0.55**

6. For the 50%-mass Bi alloy, the temperature above which the alloy is completely liquid is T=140 °C.

7. For the same alloy, the solidification temperature is T=125°C.

The eutectic composition is Xeut=56%mass-Bi.

a) The phases present at T=95°C are: β+γ.

b) Composition of each phase: Xβ=40% mass-Bi

Xγ=99.9% mass-Bi

Estimation of the mass fraction of each of these phases.

Applying the chemical moment theorem, we get :

Xβ=99.9-56/99.9-40=0.73 and Xγ= 56-40/99.9-40=0.27

8. The Pb-Bi binary equilibrium diagram is :

   A partial miscibility in the solid state.

Exercice 8

We propose to draw the liquid-vapour equilibrium diagram of the H2O-HNO3 system, under a pressure of 1 atm. To do this, we have the experimental data summarized in the following table:

Mass fraction of HNO3	0	0.1	0.15	0.2	0.3	0.4	0.5	0.6	0.66	0.7	0.75	0.8	0.85	0.9	0.95	1
T°C	100	101	102	103	105	109	115	120	123	120	114	105	99	93	88	83
T°C	100	106	109	112	116	118	120	122	123	122	120	117	112	105	96	83
	100								123							83

8. The phase diagram of the binary system H2O-HNO3.

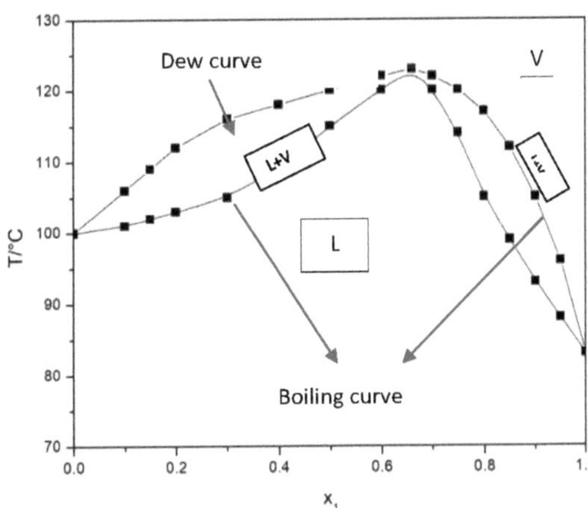

The mixture is an azeotrope indicating a negative deviation from ideality.

9. Consider a new mixture M obtained during the preparation of nitric acid. It consists of a total of 100 moles and its molar fraction of HNO3 is XHNO3=0.11.

$n_{HNO3} = x_{HNO3} * n_t = 0.11 * 100 = 11 mole$ and $n_{H20} = x_{H20} * n_t = 0.89 * 100 = 89 mole$

Hence $m_{HNO3} = n_{HNO3} * M = 11 * 63 = 693\ g$ $m_{H20} = n_{H20} * M = 89 * 18 = 1602\ g$

And $m_t = m_{HNO3} + m_{H20} = 2295\ g$

   c. fraction massique $(HNO3) = \dfrac{m_{HNO3}}{m_t} = 0.30$

The boiling point is T=105°C

10. This mixture M is maintained at 109°C.

Mass fraction of the liquid phase (0.4 in HNO3 and 0.6 in $_{H2O}$)

Mass fraction of the vapor phase (0.15 in HNO3 and 0.85 in $_{H2O}$)

Applying the chemical moment theorem we find :

Mass of the liquid phase=1377 g

Mass of the vapour phase= 918 g

CHAPTER III

Exercise 1

T/°C	x1	y1	P1= y1P	P1*	P1id=x1P1*	P1-P1id/P1id	$\ln\gamma_1$	A	P1reg	P1-P1reg/P1reg
68,2	0	0	0	2,709	0	0	0		0	0
63,1	0,108	0,231	0,231	2,352	0,254	-9,05	-0,095	-0,119	0,231	-0,18
68,2	0,310	0,539	0,539	1,832	0,568	-5,08	-0,052	-0,110	0,537	0,36
68,2	0,522	0,750	0,750	1,478	0,771	-2,78	-0,028	-0,123	0,751	-0,15
68,2	0,747	0,897	0,897	1,207	0,901	-0,50	-0,005	-0,078	0,895	0,25
68,2	1,000	1	1	1,000	1,000	0,01	-0,000		1,000	0,01

2- $\gamma_1 = \dfrac{P_1}{P_1^{id}}$ and $a = \dfrac{\ln\gamma_1}{(1-x_1)^2}$ with $a_{moyen} = -0,117$.

3- $P_1^{reg} = exp[a(1-x_1)^2]x_1P_1^*$ (the results are shown in the table above);

Exercise 2

1- $G^E = (n_1 + n_2)g^E = a\dfrac{n_1n_2}{n_1+n_2} \cdot g_j^E = \left(\dfrac{\partial G^E}{\partial n_j}\right)_{T,P,n_i \neq j}$

Hence : $g_1^E = wx_2^2$ and $g_2^E = wx_1^2$

$g_j^E = RT\ln\gamma_1$ so $\ln\gamma_1 = \dfrac{w}{RT}x_2^2$ and $\ln\gamma_2 = \dfrac{w}{RT}x_1^2$

2- $h^E = ax_1x_2$, $h^M = h^E = ax_1x_2$, $s^E = 0$ from which $s^{M,id} = S^M$

3- The relationships are used: $y_jP = \gamma_jx_jP_j^*$ to the azeotropic composition:

$$\left(\dfrac{\gamma_2}{\gamma_1}\right)_{az} = \dfrac{P_2^*}{P_1^*} \rightarrow \ln\left(\dfrac{\gamma_2}{\gamma_1}\right)_{az} = A - \dfrac{B}{T} = 3,02674 - \dfrac{778,6544}{T}$$

4- For regular mixing : $\ln\left(\dfrac{\gamma_2}{\gamma_1}\right)_{az} = \dfrac{w}{RT}(x_1^2 - x_2^2) = \dfrac{w}{RT}(x_1 - x_2) = A - \dfrac{B}{T}$

5- $x_1 = 0,66977$ et $0,71645$ à 330 K et 350 K.

Exercise 3

1- $x_1 dlna_1 + x_2 dlna_2 = 0 \rightarrow ln\gamma_2 = \frac{w}{RT} x_1^2$

2- $P_1 = \gamma_1 x_1 P_1^* = exp\left(\frac{w}{RT} x_2^2\right) P_1^*$ and $P_2 = \gamma_2 x_2 P_2^* = exp\left(\frac{w}{RT} x_1^2\right) P_2^*$

3- $w = \frac{g_1^E}{x_2^2}$ i.e. wmean =8780 J/mol (the results are summarized in the table below).

x1	0,1	0,3	0,5	0,7	0,9
w	8777	8777	8780	8777	8800
γ_1	2,4	1,7	1,31	1,7	2,4
γ_2	1,01	1,1	1,31	1,7	2,4

Bibliography

[1] Chemical thermodynamics,. Danielle Guignard. Copyright 1989. ISBN 2-7298-8969-8.

[2] Thermochemistry,. Christian Picard. De Boeck S.A,. 1998. ISBN 2-8041-2839-3.

[3] Physical and chemical thermodynamics,. Paul Roux. Edition marketing S.A, 1998. ISBN 2-7298-9887-5.

[4] Thermochemistry Binary diagrams and materials development,. Edition Bréal, 1999. ISBN 2 84291 179 2.

[5] Thermodynamic properties of nonelectrolyte solutions, William E. Acree, Jr. Academic press, INC, Copyright 1984.

[6] Chemical thermodynamics,. Martial Chabanel and Bertrand Illien. Ellipses Edition Marketing S.A., 2011. ISBN 978-2-7298-65016.

[7] Chemical thermodynamics from course to tutorial,. M.T. Achour. Office des publications universitaires 03-90.

[8] Thermodynamics of solutions and mixtures,. Jean Lozar. Ellipses Edition Marketing S.A., 2013. ISBN 978-2-7298-77972.

[9] The basics of thermodynamics,. Jean-Nöel Foussard, Edmond Julien, Stéphane Mathé and Hubert Debellefontaine. 3rd Edition Dunod, 2005, 2010, 2015. ISBN 978-2-10-072131-3.

[10] Handbook of thermodynamic diagrams,. Gulf Publishing Company, Houston, Texas. Copyright 1996.

[11] The properties of gases and liquids,. Bruce E.Poling, John M.Prausnitz, JOHN P.Oconnell. Mc Graw-hill,. USA (2001).

[12] Thermodynamic an Engineering Approach,. Yunusa Cengel, Michael A.Boles. Eighth Edition, Mc Graw-Hill USA (2011). ISBN 978-0-07-339817-4.

Printed by Books on Demand GmbH, Norderstedt / Germany